KU-598-047

SPREADSHEET GEOMECHANICS

An introduction

BURT LOOK

Materials and Geotechnical Services Branch of the Queensland Department of Transport, Queensland, Australia

624.1513
LOO

A.A.BALKEMA/ROTTERDAM/BROOKFIELD/1994

Illustration on the binding was drawn by Craig Scott.

UNIVERSITY OF STRATHCLYDE
13 SEP 1994
UNIVERSITY LIBRARY

TRADEMARKS
LOTUS 1-2-3, HAL, ALLWAYS, MANUSCRIPT, FREELANCE, IMPROV, WYSIWYG,
   VIEWER, SOLVER & BACKSOLVER are registered trademarks of Lotus Development
   Corp.
EXCEL and MS-DOS are registered trademarks of Microsoft Corp.
QUATTRO & REFLEX are registered trademarks of Borland International Inc.
HARVARD GRAPHICS is a trademark of Software Publishing Corp.
Q&A is a registered trademark of Symantec Corp.
WORDPERFECT PRESENTATIONS is a registered trademark of Workperfect Corp.
SUPERCALC is a registered trademark of Computer Associates International Inc.
DBASE is a registered trademark of Ashton-Tate Co.
RISK is a trademark of Parker Brothers.

Authorization to photocopy items for internal or personal use, or the internal or personal use of
specific clients, is granted by A.A. Balkema, Rotterdam, provided that the base fee of US$1.50 per
copy, plus US$0.10 per page is paid directly to Copyright Clearance Center, 222 Rosewood Drive,
Danvers, MA 01923, USA. For those organizations that have been granted a photocopy license by
CCC, a separate system of payment has been arranged. The fee code for users of the Transactional
Reporting Service is for the hardbound edition: 90 5410 151 2/94 US$1.50 + US$0.10, and for the
student paper edition: ISBN 90 5410 152 0/94 US$1.50 + US$0.10.

Published by
A.A. Balkema, P.O. Box 1675, 3000 BR Rotterdam, Netherlands
A.A. Balkema Publishers, Old Post Road, Brookfield, VT 05036, USA

ISBN 90 5410 151 2 hardbound edition
ISBN 90 5410 152 0 student paper edition

©1994 A.A. Balkema, Rotterdam
Printed in the Netherlands

Contents

Preface

This manual was written principally for practicing civil engineers and engineering geologists wishing to familiarise themselves with spreadsheet application tools to aid the analysis and design process. It will also be useful to postgraduate or to final year undergraduate civil engineering students who have completed soil mechanics and engineering geology courses.

Spreadsheets have been used extensively in business applications to examine alternative strategies, and have largely been ignored in the general engineering profession. This is believed to be due mainly to the lack of awareness of these easily applied tools. This manual will hopefully bridge that gap.

By way of examples in the geotechnical discipline, the manual is intended to apply the spreadsheet commands in typical analysis situations. The reader may then use the development techniques to their own work environment by learning of spreadsheet development from a technical viewpoint.

Each chapter contains a summary of the relevant theory, which provides an introduction to the subject. The examples given have been arranged in an increasing degree of difficulty in terms of the spreadsheet design but not in an increasing order of difficulty of geotechnical content. Although the examples are from the geotechnical engineering field, civil engineers would be able to apply the development techniques since the emphasis is on spreadsheet design with an application to the geomechanics area. It is hoped that the wider application of spreadsheet usage may be seen to all civil engineers and engineering geologists.

The central philosophy within the manual can be summarised as:
– A solution should ideally be in the form of graphs or tables as a calculation which produces a number does not provide sufficient information for design.
– The sensitivity of a solution should always be illustrated to provide alternative strategies. The number crunching process may be reduced considerably with the use of spreadsheets.

The spreadsheets were developed in Lotus 123 Version 2.1, but the manual was written with the added facility of Version 2.X and Version 3.X in mind. Therefore users with any of the above versions would be able to use the manual. Some basic knowledge of Lotus 123 or similar spreadsheet package is assumed.

Useful comments to an early draft of the manual was made by Tom Macbeth, Engineering Geologist at the Queensland Department of Transport. I am also grateful to my wife Gina, for her support as well as comments throughout the several drafts of the manual.

Introduction

The geotechnical engineer emerged from the cross fields of geology and civil engineering. A numerical approach to a largely interpretive field resulted. There is often not *one* unique answer to geotechnical problems but a range of solutions on which engineering judgement must then be used to rationalise the effects of sometimes competing and conflicting results.

The presentation of results in graphical and tabular format provides the mix which best illustrates a range of conditions. However the production of a range of conditions is a very time consuming and repetitive task – even after the model and parameters have been defined. Iterative procedures will be involved in all stages and the advantages of this task being done by computer are obvious.

After multiple program runs or calculations, a result table has historically been used to present the range of results. However, tables of numbers may be confusing to those not intimately involved with the detailed design process.

A graph would more easily illustrate the concept of the design, without emphasis on the numerical details. Therefore a tool is required whereby calculations can produce a range of results with the tabled results translated to a graphical form, all within the one process. Spreadsheets are one such tool.

The manual will illustrate by example, how a well set up spreadsheet can reduce the analysis process to a worksheet filling exercise, instead of the usually complicated formats of many engineering programs. There are enough complexities in a design process without having to spend valuable design time trying to understanding the format of program data entry and also having to translate the calculated results into design conditions.

Once a worksheet has been developed, the designer may then devote more time examining cost effective alternatives, rather than the necessary but time consuming number crunching progress.

It should be noted from the onset that while the designer is freed from the mechanics of calculations, the calculated result must always be validated to ensure that the result reflects the reality of the situation.

In addition to the primary objective of providing a solution range, the exercise of continuous redesign becomes less of a chore and more an exercise of refined innovation.

1

LAYOUT OF MANUAL

The emphasis will be on geotechnical engineering applications. All examples presented herein are based on actual applications used during work, and therefore reflect the type of projects handled by the author and not necessarily meant to represent the full range of applications available to the user.

The author has developed spreadsheets for highway and hydraulic applications, and the spreadsheet was found to be just as useful a tool in those application areas. This manual began by simply documenting a few of the spreadsheet programs and placing them in order of an increasing degree of difficulty and so forming sequential chapters. The geotechnical degree of difficulty may however seem out of sequence within the manual.

Chapter 1 deals with the requirements of the geotechnical engineer who must integrate a fuzzy geotechnical solution into a rigidly defined structural design approach. It considers the concept of design and the need to provide the range of conditions within the report or analysis. The purpose is best served by sensitivity analysis illustrated graphically.

NOTE: Fuzzy Logic is a term used in Expert Systems. It allows for the uncertainty in many decisions. Fuzzy sets are used for evaluating a numerical value on a spectrum of possible values.

This manual makes no attempt at being a soil mechanics text and assumes the reader has an understanding of the essential soil mechanics principles.

Chapter 2 introduces spreadsheets as a tool to be used, with the type of functions available and compares this tool with conventional programming languages.

This book is not intended to replace or substitute for any text on spreadsheet analysis since only a summary of commands are illustrated herein. It is assumed that the reader has covered a Lotus 123 tutorial or has been to a 3 day introductory Lotus course or equivalent.

The spreadsheet commands are illustrated in the subsequent chapters by using examples in geotechnical engineering. Each chapter builds on the techniques of the preceding chapters.

The layout of each chapter is as follows:
- Geotechnical models;
- Example problem;
- Spreadsheet design;
- Graphical aids;
- (Alternative spreadsheet design);
- Worksheet forms.

The spreadsheet listings are given in the appendix. The listings given should be used only as a check on the cell entries and should not be used to input line for line. The spreadsheet should be developed using the menu commands, as explained within the text.

CHAPTER 1

Spreadsheets in geotechnical engineering

Engineering problems rarely occur in text book form. Exact solutions are often desired from complex and variable problems. This is accomplished by first reducing the complexity (often by simplifying assumptions) and then using a mathematical model to predict behaviour under changing conditions. Figure 1.1 illustrates the main elements of the geotechnical process.

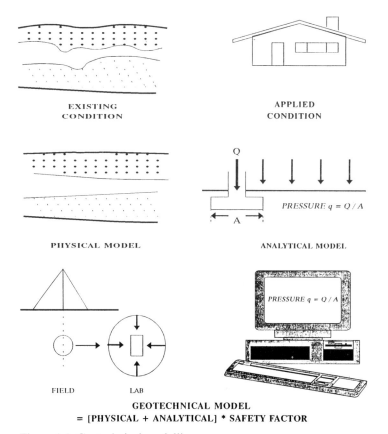

Figure 1.1. Geotechnical modelling.

A model which reduces a complex and variable problem to 'one number' is a self defeating solution. Tools must be used to provide variable solutions to variable problems. Spreadsheets are useful in this aspect.

While in a perfect world the design follows only after the design framework and plan have been formed, this is not the case in the real world of design and planning projects. This is even more pronounced in Fast Track programs where planning, design and construction sometimes become integrated. The spreadsheet can assist considerably in this iterative process.

This chapter illustrates the concept of the geotechnical modelling and the use of spreadsheets to provide a solution range which allows for the inevitable variability of data and the modelling process.

1.1 ASSESSMENT OF SOIL PROPERTIES

Geotechnical engineering is the branch of engineering which relates to the behaviour of the ground in its natural state and when some external condition is applied to it. The geotechnical engineer must assess the soil and rock properties in order to construct a physical and mathematical model of the mechanical behaviour of the material.

Soil mechanics deals with the properties of soils, its stress strain properties with time and alteration with its changing environment. Unfortunately, the determination of the actual soil properties of most sites are near impossible. The actual determined characteristics are the soil properties of selected samples of a site, and the properties within the site are expected to vary. Only the rare soil deposit approaches homogeneity.

Deductions are made with samples from boreholes which are some distance apart and typically 50 mm to 100 mm in size. The statistical implications are readily evident. Subsurface conditions are generally inferred through interpolation between borehole locations.

To compound the inference problem, the dilemma of obtaining representative samples and testing these samples provide further variability. Analysis of the physical conditions involve uncertainty of:
  – Homogeneity of the soil,
  – Disturbance, sampling and measurement errors,
  – Paucity of tests upon which to base a confident analysis.
Based on these variables a geotechnical report is usually prepared to produce the design number.

EXAMPLE: A quick undrained test provides the following values of the cohesive strength: 150, 185, 130, 215 kPa. An analysis may specify the undrained cohesion as:
  – 'Average' value = 170 kPa,
  – 'Lowest' value = 130 kPa,
  – 'Lower Quartile' value = 150 kPa,
  – etc.

> Calculations using a spreadsheet can proceed using all of the above 'design' values simultaneously, without the tediousness of constantly recalculating for the various design conditions.
> Some elementary statistical functions are also available to analyse data.

A design based on soil variables rather than a single value provides a clearer assessment of the sensitivity of the design to soil properties.

The analysis (and design) process may be automated to cover the wide range of variables to be considered. It is in this area that computers have been mainly used. Smith (1988) discusses the limitations of computers in the design process and points out that computers in the geotechnical field are presently used mainly in the automation mode rather than a design-driving mode.

1.2 THE CIVIL ENGINEERING FACTOR

The main unknowns (variables) in design would be:
- Loading conditions,
- Soil conditions,
- Construction aspects,
- Models used in design.

The geometrical parameters may be controlled to some extent and is part of the engineering design.

In geomechanics there is a physical model (an interpretative process) and the analytical model (the application of a best fit mathematical approach). This modelling process is illustrated in Figures 1.1 and 1.2.

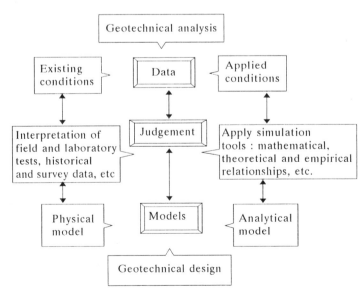

Figure 1.2. Geotechnical analysis and design.

The mathematical geotechnical models are becoming more and more refined but the main input parameters of loading and subsurface conditions are still largely interpretive. The computing industry refers to such phenomena as 'garbage in garbage out'.

Soil is a variable material. Unlike steel and concrete, the properties are not as easily controlled. To compound the inference problem the civil engineering factor would include:

– Funding of the geotechnical investigation. The geotechnical investigation represents a minor percentage of costs, yet the cost is often debated upon. The geotechnical program does not 'ADD' to the site, while the client may see plans or a physical model from the architect or civil and structural engineer as well as a physical shaping of the ideas from the contractor or builder.

– The geotechnical investigation is often placed on a low priority and sometimes requested simply 'to facilitate council requirements'. The design often proceeds before the geotechnical report is complete.

– Even after the geotechnical report has been completed, the implications are often poorly understood and poorly integrated into the design process.

The structural engineer and civil engineer often require a design number at the end of the day. The use of a number is a defeating concept to the geotechnical engineer dealing with a material which varies with all four dimensions – length, width, depth and time. The presence of water seems to sometimes provide a 'fifth dimension' to the problem and dilute the best of theoretical models.

Aside from the variabilities described, there is the daily dilemma of a design being expected to be produced without the final concept plan, levels, geometries, etc. When the geotechnical information is provided the concept plan is then altered and the geotechnical design is expected to be altered immediately – the mechanics of the analysis must then proceed for the changed condition. The spreadsheet would make this iteration within the design cycle a less mechanical process and automate the redesign in many instances – if the spreadsheet had been set up properly from the onset.

1.3 THE SAFETY FACTOR

The geotechnical engineer has to safeguard against the legal wolves waiting to strike and the professional requirement to provide as much detail as possible to the structural engineer. To account for variations and unknowns a factor of safety is often used.

Large factors of safety are applied in the geotechnical design, while in the structural design the safety factors are considerably less. The maturing of the science of geomechanics has not produced a reduction in these safety factors, and is an indication of the variability and inference involved in any geotechnical model.

Table 1.1. Typical safety factors (based on Meyerhof, 1970).

Type of design	Typical safety factor	Approximate failure frequency
Earthworks	1.3-1.5	$\dfrac{1}{500}$
Retaining structures	1.5-2.0	$\dfrac{1}{1500}$
Foundations	2.0-3.0	$\dfrac{1}{5000}$

Much is made of the factor of safety concept. However unless this is taken with the whole project in mind it may not clearly reflect the intended conditions. The factor of safety above 1.0 does not indicate that failure cannot occur. Detailed analysis may involve probability theory in order to reflect this anomaly.

Compatibility between the structural and geotechnical designs using limit state procedures involves the use of partial safety factors.

A safety factor is therefore not an end to itself but just a convenient term of reference and one of many components within the design process. The safety factor is not a number to feel 'safe' about since it represents a factor of ignorance and attempts to account for some of the anticipated variabilities of soil conditions, water levels, geometries, external loadings, etc. which will occur over a very short distance.

An engineer is not expected to refer to a factor of ignorance when discussing a project with a non engineering client or the public at large. The result is often a public made to feel 'safe' because of safety factors rather than ignorance factors. This often leads to complacency with cuts in funding and research because of the perception of continuous safety. The parameters used in design are randomly distributed rather than deterministic and hence the assignment of a number to assess an acceptable level of safety is not an absolute condition and there will always be some probability of failure as shown in Table 1.1.

However while probabilistic procedures enable the designer to assess the effect of uncertainty, a more elaborate analysis is usually required. The spreadsheet provides a commercial alternative whereby the sensitivity of the model to soil, loading or geometric variations in conditions is assessed. Some limited statistical functions are also available in the spreadsheet.

1.4 THE DESIGN RISK

In purely economic terms, the level of the risk may be examined under the broad categories of:

1. Cost of life,

2. Cost of damage,

3. Losses through a facility being out of service.

Determination of an acceptable level of risk would involve individual and social perceptions, political compromise, economic pressures, media interest, etc.

However the engineer is not an economist, but has made some allowances in the design to reduce risk. Use of a factor of safety provides the simplest compromise to allow for variables and unknowns. However a safety factor does not mean that a design cannot fail, and would vary depending on the type (importance) of the construction.

An optimisation design (cost/effectiveness) often calls for large volumes of calculations. Manual attempts at these calculations will often result in an enforced reduction in the volume to be attempted with the resulting loss of accuracy and completeness of the investigation.

Design risk curves (such as variation of factors of safety with soil properties) provide an indicator upon which sound engineering judgement may be applied and the appropriate design risk applied. The volume component of the calculation is undertaken by the program and so releasing the design engineer to consider in greater detail the complexities of the problems. Arithmetic errors are also reduced in a formal computational design approach.

A range of conditions would provide the designer with a better feel of the problem rather than a factor of safety. Graphs would be the main tool for illustration and within a spreadsheet analysis the graphs are produced automatically from the analysis results.

The spreadsheet's asset is the ease at which the sensitivity of a model to a change in parameters can be determined. This provides a very practical way to simulate the parametric uncertainty which exists in any analysis, without the use of stochastic processes.

Backanalysis of parameters can also be routinely applied to the analysis, in order to assess any changed conditions or evaluate the effects of constraints or uncertainty in a given parameter.

Again one must temper the use of numerical gymnastics to fit the real world because despite the powerful analytical tools available, it is inevitable that the final design is based not on absolutes (right versus wrong approach) but on the best judgement given the constraints of time, money and available data. While the professional engineer has the responsibility to achieve acceptable levels of safety, the safety margin may often be dictated by the funds available.

1.5 THE ENGINEERING PROGRAMMER

The geotechnical engineer comes from a range of backgrounds, which are mainly sub disciplines of civil engineering and engineering geology.

The geotechnical study may involve civil, structural or mining projects and

may involve aspects of:
- Foundation engineering,
- Soil mechanics/rock mechanics,
- Engineering seismology,
- Hydrogeology,
- Highway engineering, etc.

It is readily evident that the geotechnical engineer will be hard pressed to have a technical handle on many aspects of his broad profession. Engineers involved with the development of programming applications may find themselves one step removed from their 'true' field. However programming by engineers is a necessary evil in many instances which has already blossomed into its own speciality.

In order to maintain the emphasis on geotechnical engineering, the engineer should avail himself of the programming tools available, rather than devote the specialist time required for efficient programming.

In recent years, programs have become more user friendly using pull-down menus, window facilities, icons etc. These programs cover most, but not all, situations and a best model fit is often adapted i.e. we fit the problem to a model we can solve. With spreadsheets there is the flexibility to customise the model to better fit the true problem.

Spreadsheets are an easily applied tool which does not require a high degree of specialised knowledge of programming languages. A working geotechnical model may be developed in a matter of days or hours, depending on the complexity of the problem.

CHAPTER 2

Introduction to spreadsheets

In computer programming, conforming to the rules (grammar) of the language is known as following the language syntax. This chapter covers the syntax basics for spreadsheet development. A brief overview of Lotus 123 facilities is covered, but the reader is referred to a Lotus text or manual for detailed explanations of the spreadsheet commands.

In some applications the computer can be an asset while in others it may be a time consuming tool. The usefulness or limitations of any tool must be realised from the onset. One should not spend weeks doing by hand what can be done in hours with a computer, but one should not spend days on the computer doing what may be accomplished in minutes using a calculator.

Worksheet, graphics and database facilities are integrated in the Lotus 123 spreadsheet. However if one is concerned only with graphs then there are more comprehensive packages on the market which deal exclusively with the production of graphs and graphics facilities (such as Harvard Graphics, WordPerfect Presentations, etc.). Similarly if the spreadsheet is used only as a database then limitations readily become evident. Relational database programs include DBase, RBase, etc. and flat file data base handling programs such as Q&A, Reflex etc. The ability for file transfer exists between these packages and Lotus 123.

2.1 THE WORLD OF SPREADSHEETS

A spreadsheet is an application program. It is an analytical tool used for the development of a working model and can be used in any field of study. Worksheet, graphics and database facilities are used to develop the model.

> NOTE: Worksheets are seen one screen at a time and is the application within the spreadsheet. Spreadsheets, as used in this manual, refer to the series of worksheet screens as well as the graphs and menu commands required to develop the worksheet. The words Spreadsheet and Worksheet are sometimes used interchangeably.

A program is basically a set of instructions. In spreadsheets, the ease of programming is accomplished via a shell which is menu driven, and so aids program development, graph production, file handling as well as formatting, copying, etc. A spreadsheet may be viewed as a non-programmer's tool for program development.

The use of macros (Section 2.8) follows more 'conventional programming' methods to automate some of the processes as well as provide a user friendly interface.

> NOTE: A macro is a series of keystrokes and special commands which performs a task.

Graphs can be produced from data within the worksheet. These graphs change as data is updated in the worksheet, without the need to reset any graphical commands.

The database facilities may be used for sorting or extracting data so that statistical calculations may be performed on selected data, or graphs may be generated.

Lotus 123 version 2.1 was used to initially develop and use the worksheets. Versions 3.4 and 2.4 have improved facilities. Lotus Allways/WYSIWYG, which are add-ins to version 2.X were used for presentation effects in this manual. Because the worksheets were originally developed in version 2.1 the flexibility exists to use them in any of the mentioned versions.

The Lotus program has been used to illustrate the use of spreadsheet geomechanics simply because it is one of the best selling packages. Other programs such as Quattro Professional, Supercalc and Excel are considered superior by their users because of more facilities, greater flexibility, superior graphs, etc. However, it is not the purpose of this manual to assess spreadsheet brands as this is a fast changing and competitive environment, with the contender for the most attractive package changing every few months. All of the programs mentioned provide the integration of spreadsheets, graphics and database facilities.

The examples showed herein can be applied to any spreadsheet, but the format of the commands may vary.

This chapter provides an overview of some of the facilities available and the syntax of the spreadsheet language. Reference should be made to the relevant spreadsheet manuals or any of the wide variety of aids (books, disks, cassettes, videos) available for learning the spreadsheet commands.

Spreadsheet terms are used with an assumption of some understanding of the jargon. Therefore the reader is again referred to the Lotus manual for clarification of these terms, where they may occur.

2.2 THE WORKSHEET SCREEN

Worksheets are essentially a grid of horizontal rows and columns which can be interrelated by formulae. The cells may contain:

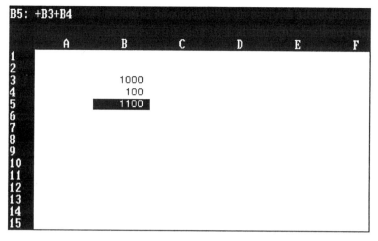

Figure 2.1. Worksheet screen with pointer at cell B5.

EXAMPLE: B5 refers to the contents of the cell referenced by the intersection of column B and row 5. The contents of cell B5 is: +B3+B4. The actual display at cell B5 is 1100, which is the result of the addition: +B3+B4 = 1000 + 100 in this example.

– text (labels),
– numbers,
– formulae, including logical and mathematical functions.

A worksheet screen is shown in Figure 2.1. It consists of a grid referenced numerically for the rows and alphabetically for the columns. If a formula is entered at a cell, then the calculated result is displayed. The input formula can be seen at the top left in the control panel area.

The control panel at the top of the worksheet screen displays the Mode Indicator, Menus (when activated), the cell address (at which the Cell Pointer is located), and its contents. Version 2.4 would have the SmartIcon to the right of the screen. The SmartIcons are a graphical representation of the 1-2-3 actions or commands.

Direction keys
Movement around the worksheet is accomplished using:

Cursor keys – across (right → : left ←) arrows
– vertical (up ↑ : down ↓) arrows
Movement – PAGE UP, PAGE DOWN
– HOME, END (with up or down keys).
GOTO key – Function key F5 to move to a specific cell.
Window – Function key F6 to move between windows if the window is split.

   – [Version 3.X] CTRL PAGE-UP or CTRL PAGE-DOWN moves the
   cell pointer between worksheets where multiple sheet files are
   in use.

2.3 COMMANDS

When you type / , a menu appears at the top of the screen. Menus are used to
execute commands. The menu contains screen prompts to aid in selection. A few
of the screen prompts are shown on a branch of the menu command tree structure
illustrated in Figure 2.2a. Depending on the version of Lotus, the command
structure may vary slightly from those shown.

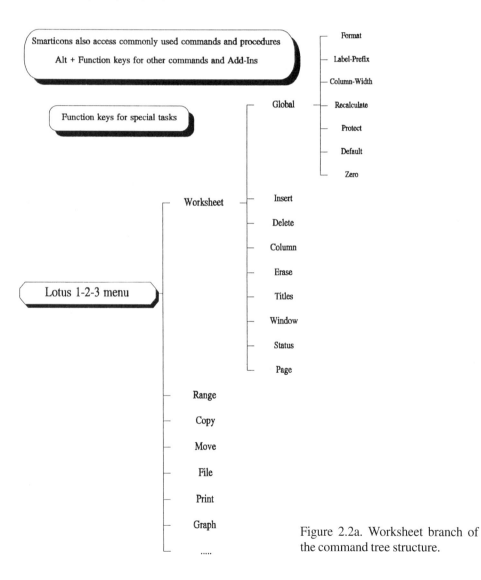

Figure 2.2a. Worksheet branch of
the command tree structure.

The user should become familiar with the command structure and sequence, as these are the building blocks upon which spreadsheet development is based.

Lotus has a unique letter assigned to each command for a given level, and each command can be invoked by using the first letter of the command.

> EXAMPLE: (Fig. 2.2a): /WGF would display the format commands (fixed, scientific, currency, etc.) and whichever option is selected will affect the entire worksheet.

The two steps in using the menu are:
1. To display the 123 menu press /,
2. Select a command by one of the movement keys or the first letter.

> EXAMPLE: To print, first activate the menu (/), and move the cursor to highlight Print or alternatively press P.

The sub-menu then appears and the command is further defined until the intended procedure is accomplished. Use the ESC key to exit from the command level or from the menu commands entirely.

Alternatively, another branch of the tree may have been chosen as shown in Figure 2.2b.

Familiarity with the Format commands will greatly enhance the presentation of the spreadsheet.

As an alternative a mouse may be used to navigate through the menu commands or with the SmartIcons in version 2.4 the selection can be made by selecting the relevant picture arranged in several palettes to the right of the worksheet.

A brief description of main menu commands are now presented.

Worksheet
Worksheet commands control the display and organisation of your work. These commands can be classified into global commands which affect the entire worksheet or specific commands which affect only certain areas.

Range
Range commands control rectangular blocks of cells. It may be used to affect specific ranges which are required to be varied from the worksheet global settings.

Copy and Move
Copy and Move commands will assist in the quick development of the worksheet by movement of selected cell ranges from one part of the worksheet to another. Copy replicates while Move transfers the range specified.

File, Print and Graph
File, Print and Graph are some of the input/output facilities. They are the

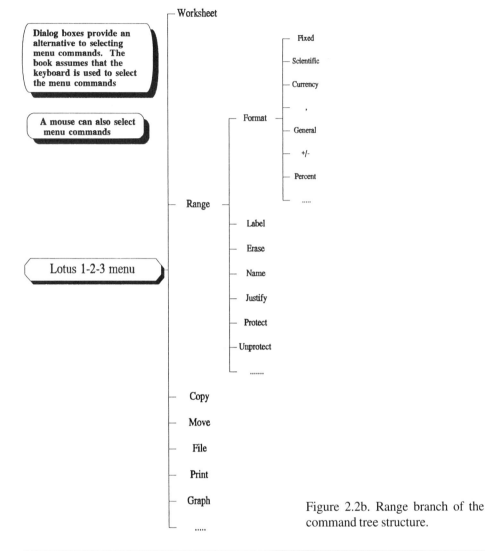

Figure 2.2b. Range branch of the command tree structure.

EXAMPLE (Fig. 2.2b): / RF to arrive at the format options for a specific range of cells. This can be compared with the global format used in Figure 2.2a where the entire worksheet would have been set to a specific format.

means by which a permanent copy of the data in the worksheet is made.

Data in the worksheet screen is temporary unless saved. File is used to Save or Retrieve a worksheet to disk, and may be in whole or part. File saving is accomplished via the menu facility as follows: / File Save {filename}.

Print outputs the specified range of the worksheet as a hardcopy or as a file for later output. Graph outputs the graph to the monitor. Graphical printing is done via the PrintGraph option after exiting 123 or using the Allways or WYSIWYG Add-Ins.

> NOTE: Version 3.X allows the printing of graphs to be completed while still within the 123 worksheet screen using / Print Image.

Data

Data is used mainly for data manipulation of a flat file database. It is also useful for entering and analysing data. There are also Matrix and Regression facilities in the Data option.

System

System allows temporary suspension of the 123 session without actually quitting from 123. To return to 123 from DOS, type Exit and press ENTER at the operating system prompt and you will be returned to the 123 session.

Quit

Quit ends the current 123 session and returns to the 123 access menu screen.

Useful commands

For the purposes of this manual a few of the more regularly used commands are briefly described.

Specific menu commands that will be useful throughout this manual would be the Window, Titles, Recalculation, Protection options.

Window allows two separate parts of the worksheet to be viewed simultaneously with the options for unsynchronised movement. This is useful for viewing large worksheets. In version 3.X the graph may be viewed simultaneously with the worksheet by using the Window menu.

Titles allows a selected row or column to be fixed since this may represent a heading for a long list of data, which requires many screen shifts.

The Recalculation is useful for large worksheets by placing the calculation on Manual and so spend less time waiting on recalculation while designing a worksheet. Once the worksheet is completed, the recalculation may be put back on its automatic mode. The Iteration feature also exists in this option.

The use of the Protect and Unprotect commands helps to safeguard certain parts of the worksheet as well as highlight the input cells. The input cells are therefore highlighted and unprotected and the formula part is protected from accidental erasures.

Release 2.3 and 2.4 provide add-ins which enhance the worksheet presentation for reporting. In version 2.4 the auditor add-in is useful in identifying relationships and dependencies, while the backsolver add-in may be used to adjust the input variable to achieve the desired end result.

2.4 1-2-3 FUNCTION KEYS

The function keys can be used to invoke 123 commands without having to use the

Table 2.1. Functions keys.

Keys	Description	Alt + Key: Description
F1	Help to use the commands	Composes international characters
F2	Edit a cell	Activates STEP mode for macros
F3	Name a range	Select macro to run
F4	Absolute value	Undo last action (if undo is enabled)
F5	Go to a cell	Learn feature to record keystrokes
F6	Change window	
F7	Query a database	Starts a 123 Add-in (if assigned)
F8	Form a table	Starts a 123 Add-in (if assigned)
F9	Re-calculate	Starts a 123 Add-in (if assigned)
F10	View a graph	Displays Add-in menu

menu or other longer procedures. They represent short-cuts in being able to define tasks more quickly.

The EDIT {F2}, ABSOLUTE {F4} and VIEW GRAPH {F10} has been used extensively in this manual. The function keys are also used with macro commands.

Use the HELP option for assistance during the development. This can be accessed by the HELP {F1} function key.

The ALT + FUNCTION keys provide further options, and may also be used to invoke add-ins.

2.5 FUNCTIONS

Functions can be logical, mathematical, financial or special functions. This is done by inserting the @ together with the function considered and the data value in brackets.

> NOTE: A function is a built-in formula that performs a specialised task. A sine function is an example of a trigonometric (mathematical) function.

The argument in brackets (X) may be a number, variable (cell reference) or mathematical expression.

> EXAMPLE: @sin(A3) represents the trigonometric sine function of the cell A3. This is given in radians.

A few of the statistical, mathematical and logical functions are given below. The reader should refer to the Lotus manual for a list of all the functions which include special, string, date and time, and financial functions.

Table 2.2. Types of functions.

@MAX(list)	Maximum value in a specified range
@ASIN(x)	Arc sine of angle x {radians}
@SIN(x)	Sine of angle x {radians}
@LOG(x)	Logarithm {base 10} of x
@PI	Number PI {3.1415926}
@IF(condition, A, B)	The logic function based on a given condition then alternative A or else condition B

2.6 CELL ENTRY

Cell entries are classified as either labels or values.

Labels

Labels are text entries which are used to provide headings, explanations, etc. However in some instances (especially in engineering applications), a 'text' description in a form may be misinterpreted in 123 as a formula. In these cases inverted commas will be required to be placed before the cell entry.

> EXAMPLE: `"Footing Width` – would be text entered into a cell. The inverted commas will not show on the worksheet screen and will be acknowledged to be text.

> EXAMPLE: `1-X1` – This will be interpreted as a formula. If it is meant to be a column heading (Text) then use an inverted comma `"1-X1`.

Text may be positioned to the left, right or centrally within a cell with the following options:

' Text left justified,

" Text right justified,

^ Text centred.

Alternatively the text may be aligned with either the global or range commands as follows:

`/ Worksheet Global Label-Prefix`

                                         {Left, Right, Centre}

`/ Worksheet Range Label`

                                         {Left, Right, Centre}

Values

Values may be numbers or formulae. Formulae are used to calculate and inter-relate cells based on numbers provided. The value may be a cell reference or an actual number. The algebra of formula cells follow conventional procedures with (brackets) and functions calculated first followed by multiplication, etc.

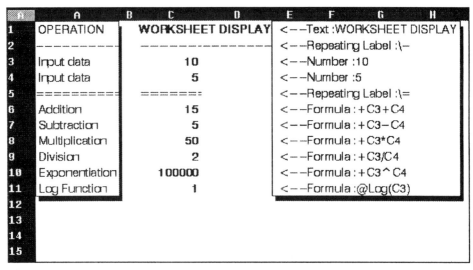

Figure 2.3. Formula, value and text cell entries.

Referencing to a cell must have the sign placed in front of the data cell.

Values are always justified to the right. Figure 2.3 provides examples of labels, values and operations used in the cell entry.

EXAMPLE: +C3 −C4 – would mean the contents of cell C4 is subtracted from cell C3.

The control panel at the top right hand corner of the worksheet will display the mode under which 123 is operating – label, value or other mode indicator.

2.7 GRAPHS

Data can often be analysed and interpreted more clearly if presented in a graph. This is used for illustration of the data calculated in the worksheet. Just as the 'numbers' in the worksheet would change based on the formula in the cell so too the graphs, once set up, would change accordingly. Numbers are often confusing and intimidating to most people while a graph would easily illustrate the range of results in a simple, easily understood format.

The facility exists for providing text labels, data labels and headers for the axis as well as the graph. Graphs may be piecharts, line graphs, bar or stacked graphs. Mixed graphs may be produced in the Lotus 3.X version, with other graph enhancements.

The option is accessed by : / Graph. The graph type, data, headings, format, etc. are then defined. The line or XY graph may be shown with symbols, lines or both. The graph is viewed by using the function key {F10} or View command.

LINE graphs are useful when the abscissa is text (1992, Height of fill, etc.).

However in engineering and sciences where the relationship between two sets of values are often required then the XY graph will be more useful.

> EXAMPLE: In defining strength variation with depth, the depths are typically input as 1.5 m, 2.25 m, 4.5 m, 7.5 m, etc. A LINE graph would plot the above figures in equal increments while the XY graph would scale the numbers to provide the correct perspective.

2.8 MACROS

Macros were originally conceived as simple keystroke capture facilities. It has since emerged with virtually full programming language capabilities. Very sophisticated automated systems can be built with these tools – when combined with the flexibility, facilities and ease of use of a spreadsheet.

Lotus can be used as a user friendly program development tool. There will be limitations, but the use of macros help bridge the shortfalls in programming facilities. Macros are used to automate and customise the spreadsheets. Multiple keystrokes may be reduced to a two keystroke operation. The advanced aspect of macro development is the command language. This controls tasks, with logic functions and customising facilities available.

At its most sophisticated level, the command language structure takes on the form of the more conventional programming languages. User friendly interfaces may also be provided to enhance and/or to assist the unfamiliar with a particular spreadsheet application. As in all programs some debugging may be required before the macros work as intended.

The use of Lotus add-in packages such as Hal provide an easier medium to create, edit or debug macros. Hal macros use English phrases rather than special representation command sequences.

Version 2.4 is now packaged with a macro library manager add-in which allows access to macros in memory rather than in a specific worksheet. Macros may be attached to the Smarticons in version 2.4 to make the application more user friendly.

2.9 WORKSHEET MANAGEMENT

Spreadsheet design provides a quick analytical model which the average engineer can quickly develop in a matter of hours, once the spreadsheet syntax is familiar.

It is highly recommended that the spreadsheet design be documented during development, as most users will find that it will be the keystone upon which modifications will be based, and often these modifications may not occur until some time after the initial spreadsheet design.

The engineer will often find that documenting the program logic and assumptions will be more time consuming than the actual program development and therefore often neglects this important aspect. In documenting the program, clearer insights will be gained into the type of assumptions which inevitably were built into the spreadsheet logic, and thus the limitations of the analysis model will be more apparent. Increased flexibility is attained by distinguishing between a value and a variable otherwise the worksheet may be limited to a job specific application.

> EXAMPLE: A 450 mm pile design usually involves calculations of the shaft and bearing resistance over the profile, and hence an output of the required pile length for that load. In this instance, the 450 mm pile diameter and the load are also variables and formulae should not include those values, but referenced to an input cell.

Planning the worksheet should occur before use of the computer. Typical steps would be:
1. Relevant theory:
   – Formulae,
   – Variables of concern,
   – Assumptions in analysis,
   – Desired output.
2. Worksheet layout mapped into the appropriate areas:
   – Data input,
   – Calculations,
   – Tabled results,
   – Macros,
   – Graph aids, etc.

As a general rule use the top left hand area of the worksheet for data input, and break the worksheet with a window so that changes from data entry are viewed simultaneously with the tabulated output without having to scroll through the worksheet.

Based on application, hardware and software, monitoring of the memory usage may be required in some large worksheets.

The file saving facility must be emphasised since it is a frustrating exercise to spend hours developing a spreadsheet only to be wiped out by some careless or unforeseen incident.

For additional facilities or enhancement of presentation, Lotus has a range of add-in packages such as Manuscript for word processing, Allways for output presentation, Freelance for business graphics, etc.

Lotus version 3.X has many improved facilities, such as easier linking of the worksheets by referencing the worksheets A, B, C, etc. This however comes at a cost to memory.

The main limitations of a spreadsheet would be the size of file that the spreadsheet can handle since Lotus 123 is RAM based. The use of this 'easy' programming shell also means that program execution time is sacrificed for the

Table 2.3. Hardware requirements of Lotus 123.

Year/Period	Version	Mb of disk space required
1983	1, 1A	0.7
1985 – 1987	2.0. 2.01	1.4
1989	3.0	4.0
1989	2.2 with Allways	3.0
1990 – 1992	3.1, 3.1⁺, 3.4	5+
1991 – 1992	2.3, 2.4	5+

benefit of user friendliness. However with powerhouse processors on the market already, and with falling prices the limitations of speed and/or RAM is not an issue for up to date hardware.

There is therefore a trade-off which must be made between relative speed and size of files as compared with the ease of programming and integrated facilities which spreadsheets can provide.

To the purists the spreadsheet will not provide the programming efficiency that is part of the conventional languages, but to the engineer who is more concerned with the application rather than the programming the use of this tool provides the compromise solution.

NOTE: On good spreadsheet design features:
1. Proper documentation. Essential for future use, whether by yourself and especially if others are to use the worksheet. Use indentifying labels for data, proper headings and notes within the worksheet.
2. Tested worksheet. Use known examples to ensure the accuracy in logic and formulae. The AUDITOR add-in analyses the worksheet to indentify circular references, dependents and precedents in formulae relationships.
3. Enhanced worksheet. This draws attention to key areas and to make bulky worksheets look less confusing. The WYSIWYG add-in allows shading, lines or boxes, bold or font adjustments.
4. Expandable worksheet. Built in flexibility which allows for future changes, such as proper usage of absolute and relative addressing and not burying 'constants' within the cell formulae. What-if-experimenting is easier when the formula arguments are placed in separate cells and then referencing those cells – this also serves to highlight the 'constants'.
5. Automatic graph linkages. What-if-graphing becomes less likely to have inconsistencies if worksheet labelling changes occur. Use cell referencing to relate the graph legends and titles to graph data.

CHAPTER 3

Stress analysis

In geotechnical engineering the variation of parameters with depth is often used in analysis. The application of spreadsheets is introduced by the analysis of stress variation with depth and offset distances for a point load condition.

The basic spreadsheet commands outlined in Chapter 2 are covered.

3.1 ELASTIC THEORY

The variation of the applied stresses within a soil mass can be determined by elastic theory. The soil mass is considered to be a semi-infinite, homogeneous, isotropic mass with a linear stress-strain relationship. The above condition holds mainly for reasonably constant soil properties to a depth of more than 5 times the width of the footing.

The induced stress at any depth is a function of distance (x, y, z) from the applied load, the magnitude, and type of applied load. A solution range is usually given by providing charts and graphs based on influence factors where the influence factor is a spatial function.

Sometimes contours of equal vertical stress are plotted to produce pressure bulbs. The stresses may be:
- Vertical stresses σ_z,
- Horizontal stresses σ_x, σ_y,
- Shear stresses τ_{xz}.

It is often assumed that the x and y horizontal stresses are equal so that $\sigma_x = \sigma_y$.

The Boussinesq equations are used herein for the stress analysis. In the Westergaard analysis, the elastic mass is considered to be reinforced laterally and stresses less than the Boussinesq values are obtained. The true condition would lie somewhere between the two extremes.

Reference can be made to standard texts such as Craig (1983) or Carter (1983) for the relevant theory.

3.2 ELEMENTS OF POINT LOAD ANALYSIS

The stress increase at a point X due to a point load Q on the surface, can be determined from the Boussinesq equations as given in Table 3.1 and illustrated in Figure 3.1.

The variation of the vertical stresses due to a point load is illustrated in Figures 3.2a and 3.2b for points directly under the point load ($r = 0$) and at specific depths (z = constant) respectively. The stress decreases with depth and offset distance from the point load.

The stresses due to more than one point load may be obtained by the principle of superposition. Surface load effects distributed over an area, line, strip, etc. is

Table 3.1. Point load stresses.

Vertical stress $\sigma_z = \dfrac{3 * Q * z^3}{2 * \pi * R^5}$

Shear stress $\tau_{xz} = \dfrac{3 * Q * r * z^2}{2 * \pi * R^5}$

where: Q = concentrated point load (kN); z = depth (metres); r = offset distance (metres) from Z axis (= $\sqrt{(x^2 + y^2)}$); x, y = offset coordinates; R = shortest distance between point load and point X (= $\sqrt{(r^2 + z^2)}$).

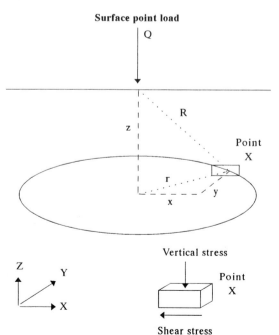

Figure 3.1. Stress due to a point load.

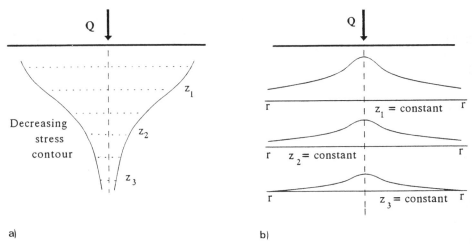

Figure 3.2a. Point load stresses ($r = 0$), b. Point load stresses at depth ($r = 0$, z = constant).

usually obtained by a modified form of the equation resulting from the integration of the Boussinesq formula.

3.3 SPREADSHEET ANALYSIS OF POINT LOADING STRESSES

PROBLEM: Calculate the variation of vertical and shear stresses with depth for a point load of 100 kN. The stresses are required for an offset distance of $x = 1$ metre and $y = 1$ metre and for the variation between 1 and 4 metres depth at every 0.5 m increment.

The theory covered will be used to develop a spreadsheet for the stated problem. This spreadsheet requires that the variation of stress with depth be shown for a given load and distance from the centerline of the point load.

A complete listing of the spreadsheet is given in the appendix and the worksheet form at the end of this chapter.

This spreadsheet will be developed in the following steps:
1. Create a data entry input form,
2. Develop the output table consisting of:
   - the table headings,
   - the depth variation.
3. Computation of:
   - the least distance (R),
   - the vertical stresses with depth,
   - the shear stresses with depth.
4. Convert the tabulated result to a graph.

WORKSHEET 3.1: STRESS VARIATION DUE TO POINT LOAD

	A	B	C	D	E	F
1	STRESS VARIATION DUE TO POINT LOAD					
2	--------	--------------	------------			
3						
4		Surface Point Load (Q) = ?			100	kN
5		Analysis begins at a depth = ?			1	metres
6		Increment depth (z) = ?			0.5	metres
7	x – Distance from centre of loading = ?				1	metres
8	y – Distance from centre of loading = ?				1	metres
9		=====>	r – Distance =		1.414	metres
10	+++++++++	++++++++++++++	+++++++++++++	+++++++++	+++++++++	
11	Depth	Least Distance	VERTICAL	SHEAR		
12	{m}	R {m}	STRESS {kPa}	STRESS {kPa}		
13	+++++++++	++++++++++++++	+++++++++++++	+++++++++	+++++++++	
14	1.00	1.73	3.06	4.33		
15	1.50	2.06	4.33	4.08		
16	2.00	2.45	4.33	3.06		
17	2.50	2.87	3.82	2.16		
18	3.00	3.32	3.21	1.51		
19	3.50	3.77	2.67	1.08		
20	4.00	4.24	2.22	0.79		

Worksheet 3.1. Display screen.

SOLUTION WORKSHEET: *Worksheet 3.1*
This shows the completed worksheet. Use this as a guide during the development. The following Lotus functions will be explained during the development:
 – Use of formula, values and text for cell entries
 – Copy for speed of development
 – EDIT using the F2 function key
 – ABSOLUTE ($) and relative addressing
 – Range formatting
 – Setting of column widths
 – Addition, subtraction, multiplication and division
 – @SQRT (X) for the square root of X
 – X^n for X raised to the power of n
 – @PI for the mathematical constant
 – Protect/Unprotect features
 – Print command for hardcopy output
 – Graph command for graphical output
 – Save command for file saving
 – Name graph for multiple graphs in the same worksheet.
It is assumed that the reader is familiar with movement across the spreadsheet as well as calling up the menu commands. Refer to Chapter 2 for a summary explanation of the above.

A step by step approach is adopted in this chapter to develop this worksheet only as the basic spreadsheet tools are covered. Following this worksheet the level of detail will not be as comprehensive as the reader would then be expected to have a grasp of the basic Lotus commands.

A lot of time is spent initially on dressing up the worksheet to make it meaningful to other users and including reminders to the developer as often one can forget the assumptions or analytical format used a few months later. All variables should be defined initially and kept in one area of the worksheet.

Step 1: Data entry input form

A data entry screen heads the start of the worksheet to define the 'known' variables as shown in the data entry form.

A	B	C	D	E	F
STRESS VARIATION DUE TO POINT LOAD					
--------	--------------	-------------			
	Surface Point Load (Q) = ?			100	kN
	Analysis begins at a depth = ?			1	metres
	Increment depth (z) = ?			1	metres
x – Distance from centre of loading = ?				1	metres
y – Distance from centre of loading = ?				1	metres
	=====>	r – Distance =		1.414	metres

Worksheet 3.1. Rows 1 to 9: Data entry input form.

Type in the entries exactly as shown in the required cells (rows 1 to 8 and columns A to F). You will find that the numeric entries (data) in column E, covers the end text of the 'X & Y distance from centre of loading' (rows 7 & 8). In order to display the full text, the column width should be adjusted.

Column width
In this worksheet the column settings have been varied to improve the readability of the worksheet. This will become more apparent as the development of the worksheet proceeds. This is accomplished by using: / Worksheet Column Set-Width.

> NOTE: A global column width may be set by using: / Worksheet Global Column-Width.
> For Lotus 3.X this command will set the global column width of the current worksheet only. Using Group mode on will adjust all worksheets in the current file.

The default setting of the spreadsheet is for a column width of 9. The width may be set by typing a number, or by using the across [← →] cursor keys to adjust the column width. The cell pointer must be in the column that is required to be changed if the cursor keys are used to expand the column. The columns B and C have been set to a width of 14 and 13 respectively. The full text is now displayed.

Edit
If any typing mistakes are made, the contents of the cell may be edited afterwards by using the {F2} EDIT function key while over the erroneous cell. Go to the error cell, press {F2} and make the required correction.

Text and numeric entries
The numerical values in column E are the data entries and are shown highlighted in the worksheet display screen to emphasise that they are the data input cells. Columns A, B, C, D and F are simply text entry.
    Text is usually aligned (justified) to the left edge of a cell unless specified otherwise. Use the SPACE BAR to shift the text to the right. The text entries will overlap into the right adjacent column if the column width is too small. This overlap will only occur where no entry exists in the right adjoining column.

EXAMPLE: Cells A1, B4, B5, B6, A7, A8 in the data entry input form shown.

    In addition, text cell entries may be aligned by using:
" Right justified,
^ Centre,
' Left justified.
Alternatively `Range` or `Global` settings may be set with the menu commands (Section 2.6).
    Numerical values are aligned to the right and cannot be changed.
    The above were numeric and text entries into the cells. The entry for a formula will now be used.

Formula entry
The offset distance (r) is calculated in the cell below the data entry values by using the simple relationship $x^2 + y^2 = r^2$. This requires the use of the Lotus square root and power functions.

Cell	Display	Cell Entry Formula	Heading
E9	1.414	@SQRT(E7^2+E8^2)	r–Distance

Worksheet 3.1. Formula for *r*-distance.

    The @ function is used for entry of any mathematical, logical, date, financial or special functions. In this instance we are defining a square root function @SQRT(..).
    Within the square root function, there are squared values. A squared value is a number raised to the power 2. The symbol ^ is used for any power function i.e. whether the power is to a value 2, 10 or otherwise.

First enter the text headings as shown in the data entry input form, row 9.

At cell E9 enter : `@SQRT(E7^2+E8^2)`.

In algebraic notation this is: $\sqrt{(x^2 + y^2)}$ where cells E7 and E8 are the X and Y distances. The cell formula may be input in two ways:

(1) Type: `@SQRT(E7^2+E8^2)` or

(2) Type: `@SQRT(`

Move cursor to E7, type: `^2+`

Move cursor to E8, type: `^2)`

ENTER

The rest of the worksheet will deal with the entry of the formula to calculate the range of stresses required.

Format command

The user will notice that the Solution Worksheet 3.1 shows the value 1.414 for the calculated r – distance. This is obtained by setting the format for the number of decimal places to 3. The default setting of the spreadsheet would be to calculate to the full width of the cells unless a format command is set either with the global or range commands.

In this example, go to cell entry E9 and use: `/ Range Format Fixed` {Number of decimal places = 3}

If this is not done, the number of decimal places will be dependent on the column width. Having an excess of decimal places often confuses the worksheet and may sometimes lead to a false sense of confidence in the degree of accuracy of the calculated numbers. Irrespective of the column width or format setting the calculations occur to 18 decimal places.

Formatting to a fixed number of decimal places will also be used in the output table.

The worksheet should be saved periodically by: `/ File Save` {filename – STRESSP}

Step 2: Output table

This will consist of:

– The table headings,
– The depth variations,
– Computation of the least distance (*R*),
– Computation of the vertical stresses with depth,
– Computation of the shear stresses with depth.

The headers must first be created as text entries. Therefore type the table headings as shown in Worksheet 3.1 for depth, least distance, vertical and shear stresses in the range A11 to D12. These text headings will require justification to display an easy to read worksheet as shown. Refer to the appendix listings to see which cell entries have been justified. Separator lines (+ + +) are used for this table of headings.

A	B	C	D	E
+++++++	++++++++++++	++++++++++	++++++++	+++++++
Depth	Least Distance	VERTICAL	SHEAR	
{m}	R {m}	STRESS {kPa}	STRESS {kPa}	
+++++++	++++++++++++	++++++++++	++++++++	+++++++

Worksheet 3.1. Output table headings.

Type in the output table headings as shown. The reason for having the B and C column widths set to 14 and 13 respectively should be now apparent.

Copy command
For speed of worksheet development the Copy command is used.

The separator line is drawn by using: \+. This repeats the + within the cell A10. This cell entry may now be copied to form the label header line.

Copying is accomplished by the commands:

/ Copy from {cell or range of cells: A10} to {starting cell – ending cell: B10.E10}

This copies the entry in cell A10 to cells B10 to E10.

This line at row 10 may now be copied to row 13.

/ Copy from A10.E10 to A13

Note that the destination was specified as only one cell entry but the line was copied along A13 to F13.

The cells can be specified by either typing the cell reference or by pointing to the cells via the cursor keys.

After entering the data entry screen and headings, as described, the formula entry for the output table of the spreadsheet can now be developed. However, before proceeding with the description of the output table development, the Copy command will be explained further.

Copy command: Absolute and relative addressing
Spreadsheets use relative addressing in copying formulae unless otherwise specified. Therefore referencing in a formula may be relative (default condition), absolute (a particular cell, column or row is set as an anchor), or mixed (relative and absolute combined).

> EXAMPLE: This spreadsheet development requires that the variation of stress with depth be analysed, but for a given loading Q. Therefore Q would be considered absolute in any copy command while the depth variation would be considered relative.

The stress formula can then be the input once and copied to the rows below. The fixed data values specified at the start of the spreadsheet must be made absolute when referenced, since the `Copy` command is invoked relative to the cell specified in the formula.

> EXAMPLE: The cell E4 (Point Load Q) is kept constant by pressing the {F4} ABSOLUTE function key during the entry of the formula into the cell C14. Thus when the formula is copied, the depths (relative values) would change but the point load Q (absolute value) would not vary.

Cell references within the formula can be made ABSOLUTE by the function key {F4} or the use of a $ sign. The row, column or both (i.e. the cell) may be made absolute. An absolute element is specified by a $ sign in front of the referencing cell letter and/or number. The referencing format is as follows:
- $E$6: Element E6 is made absolute,
- $E6: Column E is made absolute but row 6 may vary,
- E$6: Row 6 is made absolute but column E may vary.

The formulae for stress variation given in Table 3.1 are therefore input into the cell elements by using the `Copy` with the ABSOLUTE commands.

Output table formulae
The output table formulae is developed by:
- Specifying the depth range in column A, by reference to the initial depth and

Cell	Display	Cell Entry Formula	Heading/Formula
A14	1.00	+E5	Depth
A15	1.50	+A14+$E$6	(Increasing depth)
A16	2.00	+A15+$E$6	
A17	2.50	+A16+$E$6	
		etc.	
B14	1.73	@SQRT(+A14^2+$E$9^2)	Least distance
B15	2.06	@SQRT(+A15^2+$E$9^2)	$R = \sqrt{(r^2+z^2)}$
B16	2.45	@SQRT(+A16^2+$E$9^2)	
B17	2.87	@SQRT(+A17^2+$E$9^2)	
		etc.	
C14	3.06	3*$E$4*A14^3/(2*@PI*B14^5)	Vertical Stress
C15	4.33	3*$E$4*A15^3/(2*@PI*B15^5)	$\sigma_z = 3Qz^3 / 2\pi R^5$
C16	4.33	3*$E$4*A16^3/(2*@PI*B16^5)	
C17	3.82	3*$E$4*A17^3/(2*@PI*B17^5)	
		etc.	
D14	4.33	3*$E$4*$E$9*A14^2/(2*@PI*B14^5)	Shear Stress
D15	4.08	3*$E$4*$E$9*A15^2/(2*@PI*B15^5)	$\tau_{rz} = 3Qrz^2 / 2\pi R^5$
D16	3.06	3*$E$4*$E$9*A16^2/(2*@PI*B16^5)	
D17	2.16	3*$E$4*$E$9*A17^2/(2*@PI*B17^5)	
		etc.	

Worksheet 3.1. Formula entry for output table.

increment depth. The stress formula given in Table 3.1 would be placed in to the adjacent columns.

 – The stress variation with depth is calculated by reference to the depths specified in column A and the input data of load (Q) and distance (r).

A column: Depth
1. Move the cursor to cell A14.
2. At cell A14 enter: +E5.
3. Move the cursor to cell E15.
4. At cell A15 enter: +A14+$E$6.
5. Enter the menu commands: / Copy from [ENTER] – A15 then specified, to – A16.A20.

The contents of cell A15 are then copied to cells A16 to A20 by adding the contents of cell E6 {depth increment} to the cell directly above. The depth increment is specified as an ABSOLUTE value ($E$6). Use the F4 function key or type $.

It should be noted that the values actually displayed in column A is a function of the initial depth from surface (Reference cell E5) and the increment depth (Reference cell E6).

B column: Least distance R
1. Move the cursor to cell B14.
2. At cell B14 enter: @SQRT(A14^2+$E$9^2).
3. Enter the menu commands: / Copy from [ENTER] – B14 then specified, to – B15.B20.

The contents of cell B14 are then copied to cells B15 to B20 by changing the depth (contents of the corresponding A cell) each time, while keeping r – distance constant {absolute value}.

In algebraic notation this is $R = \sqrt{(z^2 + r^2)}$ where cells A14 and E9 are the z distance and r distance respectively (Figure 3.1).

C column: Vertical stress (Table 3.1)
1. Move the cursor to cell C14.
2. At cell C14 enter: 3*$E$4*A14^3/(2*@pi*B14^5).
3. Enter the menu commands: / Copy from {ENTER} – C14, to – C15.C20.
The contents of cell C14 are then replicated relative to columns A & B while keeping the value at cell E4 as constant.

Note the use of the @PI mathematical function which specifies the π constant. The use of multiplication *, division /, and mathematical functions should also be noted. The order of calculation follows standard algebraic format with functions and enclosed brackets computed first, followed by multiplication or division and then addition or subtraction.

D column: Shear stress (Table 3.1)
As for C column but using the formula for shear stresses.

The formula result is shown in the Worksheet Form 3.1 and not the formula entry. Both the result and the actual cell entries are given in the output table shown. The actual output table would show only the already computed results.

The worksheet output table was formatted for 2 decimal places. This is obtained by setting the format for the number of decimal places to 2, otherwise the default setting of the spreadsheet would apply. This is obtained by:

`/ Range {A14.D20} Format Fixed {2 decimal places}`

The worksheet is now complete in its development. The user may now enhance or protect the worksheet or produce a graphical output or printed copy.

Protect/Unprotect
The program allows the cells where an input is required to be highlighted. This also acts as a safeguard against accidental erasures by having the formula cells protected while unprotecting the data cells. This is accomplished by first protecting the entire spreadsheet by the following menu commands:

`/ Worksheet Global Protection Enable`

The data entry cells to be unprotected and highlighted are then referenced by:

`/ Range Unprotect {point or enter cell range: E4.E8}`

Worksheet Form 3.1 shows these highlighted input cells in bold.

A colour monitor will display two different colours for the protected and unprotected cells. A monochrome monitor will display the cells at different contrasts.

If any editing of the worksheet is required (besides the unprotected cells), then the `Protection` will have to be disabled.

`/ Worksheet Global Protection Disable`

Printing
The entire worksheet may be printed by using the `Print` options as follows:

`/ Print Printer Range A1.F20 Go`

Reference should be made to the Lotus manual for further descriptions on the use of the `Print` options, such as specifying a header or footer, margins, etc.

NOTE: Lotus 3.X has other print options such as printing with the borders (full frame labelling in terms of rows and columns), sideways printing (portrait or landscape), fonts change, standard/compressed, etc. Lotus 2.X can accomplish these tasks with add-in packages or printer commands.

Step 3: Graphical output

The calculated table can be illustrated by a graph. The graph would typically be plotted by having the stresses (shear and vertical) shown varying with depth. This graph is an X-Y graph. The graph command is invoked by:

```
/ Graph Type X-Y
              X (A14.A20)                          ... Depth
              A (C14.C20)                   ... Vertical stress
              B (D14.D20)                     ... Shear stress.
```

The graph titles can also be set and other graph specifics, such as legend, titles, grid, format, etc. Any heading may be specified by a text or a cell reference.

```
/ Graph Options Legend
              A - Vertical stress
              B - Shear stress
```

```
/ Graph Options Titles
              First {heading on graph}      Point load stresses
              Second {subheading}                     Project X
              X - axis                          Depth - metres
              Y - axis                            Stress - kPa
```

Quit or ESC to move out of Options.
    While in the graph menu the graph may be viewed by:

```
/ Graph View.
```

In the worksheet (without the menu displayed) the current graph may be viewed by pressing the {F10} function key.
    The graphical translation of the tabulated results in Worksheet 3.1 is shown in Graphsheet 3.1.
    To display/remove the grid use:

```
/ Graph Options Grid Both (to see grid)
                    Clear (to remove)
```

In order to print the graph, it must first be saved by:

```
/ Graph Save {filename - Stressp}.
```

The graph is subsequently printed by Quiting 123 and using the Print Graph function.

NOTE: Lotus 3.X has the facility to PRINT while still in 123. This is done by: / Print Printer Image Current-Graph Go.

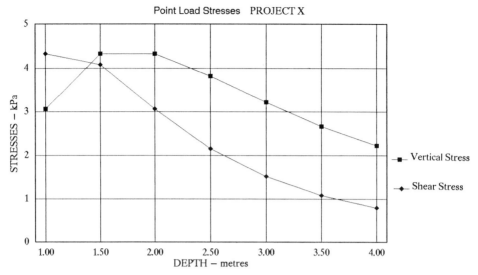

Graphsheet 3.1. Stresses with depth.

WORKSHEET 3.2: STRESS VARIATION DUE TO POINT LOAD WITH VARYING OFFSET DISTANCES

	A	B	C	D	E	F	G
1	STRESS VARIATION DUE TO POINT LOAD						
2	---------	-----------	-------				
3	Surface Point Load (Q) = ?				100	kN	
4	Initial depth from surface = ?				1	metres	
5	Increment depth (z) = ?				0.5	metres	
6	x – Distance from centre of loading				1	metres	
7	y – Distance from centre of loading				1	metres	
8	Offset distance (r) = ?				1.414	metres	
9	Offset increment (dr) = ?				0.5	metres	
10	+++++++++	+++++++++++	++++++++	++++++++	++++++	++++++	++++++++
11	Offset r–>	0.9	1.4	1.9	0.9	1.4	1.9
12	Depth z	VERTICAL STRESS {kPa}			SHEAR STRESS {kPa}		
13	+++++++++	+++++++++++	++++++++	++++++++	++++++	++++++	++++++++
14	1.0	10.46	3.06	1.02	9.56	4.33	1.95
15	1.5	9.63	4.33	1.89	5.87	4.08	2.42
16	2.0	7.43	4.33	2.35	3.40	3.06	2.25
17	2.5	5.58	3.82	2.41	2.04	2.16	1.85
18	3.0	4.25	3.21	2.26	1.29	1.51	1.44
19	3.5	3.30	2.67	2.03	0.86	1.08	1.11
20	4.0	2.63	2.22	1.78	0.60	0.79	0.85
21							

Worksheet 3.2. Display screen.

For a different approach, the spreadsheet may be developed to provide a wider range of information, such as the variation of stresses with offset distances.

In Worksheet 3.1 a solution was obtained for a specific offset distance ($X = 1$,

Y = 1). However, one may may wish to evaluate the sensitivity of the model to the specified offset distance. This would provide greater flexibility in a parametric evaluation, by incorporating the variation of vertical and shear stresses as well as the variation of the stresses with offset distances.

Worksheet and Graphsheet 3.2 show the variation of stresses with 1 offset increment on either side from the specified offset distance (X = 1, Y = 1 resulting in r = 1.414). The offset increment in this case is specified as 0.5 metres.

Cell	Display	Cell Entry Formula	Heading/Formula
			VERTICAL STRESS {kPa}
B11	0.9	+E8−E9	Offset : r−dr
C11	1.4	+E8	Offset : r
D11	1.9	+E8+E9	Offset : r+dr
			SHEAR STRESS {kPa}
E11	0.9	+E8−E9	Offset : r−dr
F11	1.4	+E8	Offset : r
G11	1.9	+E8+E9	Offset : r+dr

Worksheet 3.2. Output table headings.

This worksheet development follows from the previous spreadsheet and the user should be able to modify Worksheet 3.1 to show the changes.

The main difference between the worksheets would occur at the output table headings. The common 'R' distance was not able to be accommodated into a specific column because of the additional need to have the variation from offset distance r. The R distance formula was therefore incorporated into the cell entries under each intersection point of depth and offset distance.

Cell	Display	Cell Entry Formula	Heading/Formula
A14	1.00	+E5	Depth
A15	1.50	+A14+$E$6	(Increasing depth)
A16	2.00	+A15+$E$6	
A17	2.50	+A16+$E$6	
		etc.	
B14	10.46	3*$E$3*A14^3/(2*@PI*($B$11^2+A14^2)^(5/2))	Vertical Stress
B15	9.63	3*$E$3*A15^3/(2*@PI*($B$11^2+A15^2)^(5/2))	(0.9 m offset)
B16	7.43	3*$E$3*A16^3/(2*@PI*($B$11^2+A16^2)^(5/2))	
		etc.	

Worksheet 3.2. Output table: Formula entry.

Only a brief description of the differences between the two worksheets is given.

The headings display 1 increment distance to either side of the specified offset distance (E8 = 1.4). This is done with both the vertical and shear stresses.

The R distance $\{\sqrt{(r^2 + z^2)}\}$ is therefore incorporated into the cell entry formula without the need to create a separate column.

EXAMPLE: In cell B14:
```
3*$E$3*A14^3/(2*@PI*($B$11^2+A14^2)^(5/2)).
```
The R^2 distance is given by: `$B$11^2+A14^2`.

The user should also note the formatting of the worksheet by setting the column widths and the number of decimal places for the calculated results. This is shown in the complete listing of the cell entries given in the appendix for file 'Stressq'.

The graph settings are also similar to Graphsheet 3.1, but with the full range specified (A....F). the first 3 points plotted are for the vertical stresses for offset distances 0.9, 1.4, 1.9 in ranges A, B and C respectively. The other 3 points are for the shear stresses also for offset distances 0.9, 1.4, 1.9 and in ranges D, E and F respectively.

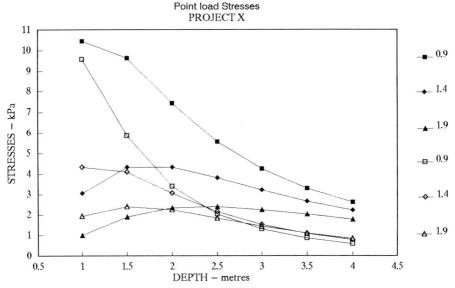

Graphsheet 3.2. Combined vertical and shear stresses.

The following graph commands are used:

```
/ Graph Type X-Y
            X (A14.A20)                    ... Depth
            A (B14.B20)          ... Vertical stress at 0.9 m offset
            B (C14.C20)          ... Vertical stress at 1.4 m offset
            C (D14.D20)          ... Vertical stress at 1.9 m offset

            D (E14.E20)            ... Shear stress at 0.9 m offset
```

E (F14.F20)   ... Shear stress at 1.4 m offset
F (G14.G20)   ... Shear stress at 1.9 m offset

The graph titles and format are set as previously indicated for Graphsheet 3.1.

The graphsheet could be separated into 2 graphs for clarity by using shear stresses and vertical stresses as the separate ordinate.

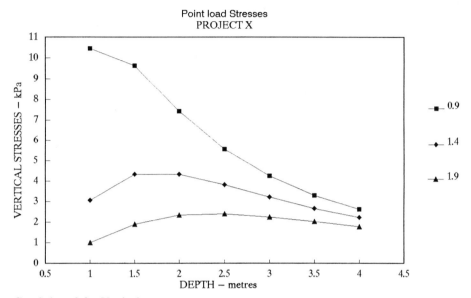

Graphsheet 3.3a. Vertical stresses.

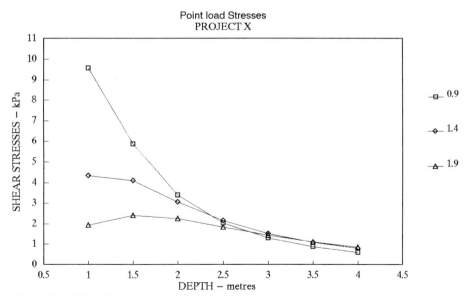

Graphsheet 3.3b. Shear stresses.

First Name the graph

/ Graph Name Create   ... {Combined}

Then adapt the previous settings using:

/ Graph Reset
        D
        E
        F

The range settings for the shear stresses have then been deleted. Change the graph Y-axis title to – Vertical stresses. Graphsheet 3.3a is produced. Name this graph Vertical.

Repeat the above for the shear stress graph (Graphsheet 3.3b) and Name – Shear.

To use any of the named graphs:

/ Graph Name Use

It will be seen that the stresses decrease rapidly with depth and offset distance for a point load. The shear stress directly below the point load ($r = 0$) is zero but increases rapidly immediately adjacent to the centreline of the load to a maximum value, then decreases rapidly thereafter. In contrast, the vertical stress immediately under the point load at $z = 0$ gives a very high value, which would produce an error in the cell (an indeterminate value). The upper bound value would be the magnitude of the imposed load.

This alternative spreadsheet shows that the structure of the worksheet can be modified to suit the variabilities that require detailed examination.

```
STRESS VARIATION DUE TO POINT LOAD
---------------------------------

                Surface Point Load ( Q ) = ?                    100 kN
               Analysis begins at a depth = ?                    1 metres
                 Increment depth ( z ) = ?                      0.5 metres
   x − Distance from centre of loading = ?                       1 metres
   y − Distance from centre of loading = ?                       1 metres
               =====> r − Distance =                          1.414 metres
```

Depth {m}	Least Distance R {m}	VERTICAL STRESS {kPa}	SHEAR STRESS {kPa}
1.00	1.73	3.06	4.33
1.50	2.06	4.33	4.08
2.00	2.45	4.33	3.06
2.50	2.87	3.82	2.16
3.00	3.32	3.21	1.51
3.50	3.77	2.67	1.08
4.00	4.24	2.22	0.79

Worksheet form 3.1. Data entry and output table. Filename: Stressp.

```
STRESS VARIATION DUE TO POINT LOAD
---------------------------------

   Surface Point Load ( Q ) = ?                  100 kN
   Initial depth from surface = ?                  1 metres
     Increment depth ( z ) = ?                    0.5 metres
   x − Distance from centre of loading            1 metres
   y − Distance from centre of loading            1 metres
            Offset distance (r) = ?            1.414 metres
            Offset increment (dr) = ?             0.5 metres
```

Offset r−>	0.9	1.4	1.9	0.9	1.4	1.9
Depth z	VERTICAL STRESS {kPa}			SHEAR STRESS {kPa}		
1.0	10.46	3.06	1.02	9.56	4.33	1.95
1.5	9.63	4.33	1.89	5.87	4.08	2.42
2.0	7.43	4.33	2.35	3.40	3.06	2.25
2.5	5.58	3.82	2.41	2.04	2.16	1.85
3.0	4.25	3.21	2.26	1.29	1.51	1.44
3.5	3.30	2.67	2.03	0.86	1.08	1.11
4.0	2.63	2.22	1.78	0.60	0.79	0.85

Worksheet form 3.2. Data entry and output table. Filename: Stressq.

CHAPTER 4

Wall analysis

When elastic stresses behind a retaining wall are required then a modified Boussinesq equation is used. The ground behind a retaining wall exerts lateral pressures in addition to the elastic stresses covered in Chapter 3. The overall wall pressure is dependent on the method of construction, soils behind the wall, movement experienced, surcharge loading and drainage conditions. The relative magnitudes of these would require to be examined to determine the critical conditions.

The basic spreadsheet commands given in Chapter 3 are expanded on. The use of multiple analysis in the same spreadsheet is used in the wall analysis and logic functions are introduced.

4.1 WALL STRESSES

A surface line or strip load will produce a stress increase in the ground. This may be calculated from an integrated Boussinesq solution for vertical, lateral (X & Y) or shear stresses. However when a wall is placed so that lateral support is no longer available then the stress equation requires modification.

> NOTE: The ground will experience STRESSES (an internal condition), while the wall will experience the PRESSURES (an external condition) induced from these STRESSES.

A wall may also experience lateral earth pressures, water pressures, as well as the effect from compaction induced effects if compaction or vibrating loads occur adjacent to the wall. The stress state is dependent on the movement experienced.

This chapter will show the development of a spreadsheet for a wall analysis with the following conditions:
- Line load surcharge (based on elastic analysis),
- Earth pressures (based on a movement criteria),
- Compaction induced stresses (based on compaction effects).
In this case where there is a wide variety of interactive forces then the spreadsheet

is designed to calculate the relative magnitudes, thereby identifying design conditions.

Reference can be made to Geoguide 1 (1982) or NAVFAC (1986) for the relevant theory.

4.2 LINE LOAD SURCHARGE ON WALL

The elastic stresses induced behind a retaining wall with a line load at the top is calculated herein using the modified Boussinesq equation. This condition is illustrated in Figure 4.1, with the cumulative pressure distribution along the height of the wall and the resulting thrust on the wall.

The recommended values for calculation of these pressures induced against rigid retaining walls are given in Table 4.1 and the resulting thrust in Table 4.2. The equations show that the formula applied in the analysis will be dependent on the offset distance of the line load from the wall.

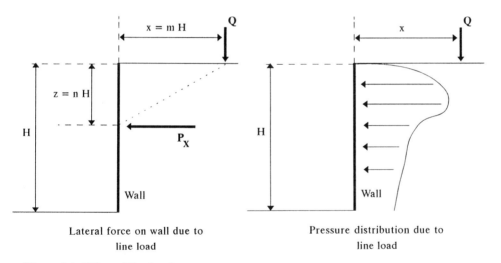

Lateral force on wall due to
line load

Pressure distribution due to
line load

Figure 4.1. Effect of line load.

Table 4.1. Horizontal stresses due to line load above wall.

$$\sigma_h = \frac{4 * Q_L}{\pi * H} * \frac{m^2 * n}{[m^2 + n^2]^2}, \quad m > 0.4$$

$$\sigma_h = \frac{.203 * Q_L}{H} * \frac{n}{[0.16 + n^2]^2}, \quad m \le 0.4$$

where: σ_h = Horizontal stresses; m = Ratio of distance (x) from wall to wall height (H); n = Ratio of depth (z) along wall to the wall height (H); P_x = Total thrust.

Table 4.2. Horizontal thrust due to line load above wall.

$$P_x = \frac{2 * Q_L}{\pi} * \frac{1}{m^2 + 1}, m > 0.4$$

$$P_x = 0.548 * Q_L, m \leq 0.4$$

Table 4.2 shows that the total resulting thrust is independent of the height of the wall (except through the m ratio). However the shape of the pressure distribution is required to provide the location of the point of maximum pressure in order to design the wall for factors such as position of tie back anchors and overturning moments.

The shape of the pressure distribution is often compared with an equivalent pressure distribution diagram. The equivalent diagram is the ratio of total thrust on the wall to the wall height.

4.3 EARTH PRESSURES

The extent of the movement of the wall for the type of soil will determine whether the at rest or active conditions exist behind the wall (Figure 4.2). This takes the form of a triangular pressure distribution for long term conditions.

The earth pressure coefficient of active, at rest or passive condition will depend on the direction and magnitude of the wall movement. The active and passive states are considered limiting conditions and represent the maximum values the wall will experience from the stress state. If insufficient movement occurs to mobilise the active/passive states then the earth pressure coefficient value lies between the at rest state and limiting value.

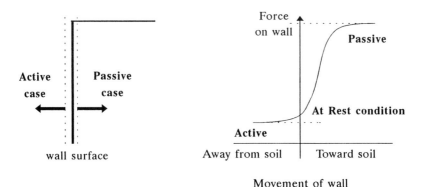

Active and Passive States

Figure 4.2. Wall movements and stress state.

Table 4.3. Earth pressures.

$$P = \frac{1}{2} * \gamma * H * K$$

where: γ = unit weight of the soil; H = height of wall; K = earth pressure coefficient.

Table 4.4. Earth pressure coefficients.

$$K_o = 1 - \sin \phi$$

$$K_a = \frac{1}{K_p} = \frac{1 - \sin \phi}{1 + \sin \phi}$$

The earth pressure P is given by: Table 4.3.
The earth pressure coefficient may be determined from the angle of friction (ϕ). The at rest pressure coefficient (K_o), active (K_a) and passive (K_p) earth pressure coefficients are given by: Table 4.4.

Water pressures behind the wall or surcharge placed at the top of the wall also provide a significant effect on the wall pressures.

4.4 COMPACTION INDUCED PRESSURES

Besides the active/passive and induced elastic stresses from surface loading, pressures may be also induced by compaction, vibration, etc. It is useful to show their relative magnitude and significance to the design.

Various theories hold as to how the wall will experience the compaction induced effect. Two methods of analysis are:

1. A modified at rest pressure coefficient K_{oc},
2. A compaction induced high level stress distribution which meets the active condition at a calculated depth.

The modified at rest pressure induced by compaction K_{oc} is given by (Perason-Kirk 1976): Table 4.5.
However method 2 for a compaction induced effect has gained more general acceptance and is illustrated in Figure 4.3 (Ingold 1979).

Table 4.5. Modified at rest pressure coefficient.

$$K_{oc} = K_o + 0.025 * (\gamma_c - \gamma_L)$$

where; γ_c = compacted density (kg/m³); γ_L = minimum (low) density.

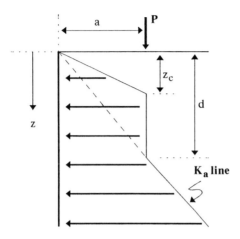

Figure 4.3. Compaction induced pressures.

Table 4.6. Horizontal stresses due to compaction effort.

$$\sigma_H = \sqrt{\frac{2 * P * \gamma}{\pi}} * \frac{L}{x + L}, \quad Z_c \leq z \leq d$$

$$\sigma_H = K_a * \gamma * z, \quad z > d$$

Table 4.7. Change point depths.

$$z_c = k_a * \sqrt{\frac{2 * P}{\pi * \gamma}}$$

$$d = \frac{1}{k_a} * \sqrt{\frac{2 * P}{\pi * \gamma}}$$

where: z = depth variable; x = distance of roller from wall = $m * H$; L = length of roller;

P = equivalent line load due to roller = $\dfrac{\text{Width of roller}}{\text{Dead weight of roller + centrifugal force}}$

k_a = active earth pressure coefficient.

The critical depth (z_c) and influence depth (d) represent change points in the analysis. The induced compaction pressures occur at the upper part of the area behind the wall and meet the active earth pressure line at a depth – d, where the effect is no longer felt.

The actual maximum value of the horizontal pressures at any point resulting from the compaction effort of the moving load is dependent on the magnitude of the roller load, the length of the roller and the offset distance behind the wall.

The line load effects, earth pressure theory, and compaction induced effects on wall stresses will now be applied to develop the spreadsheet for the stated problem.

4.5 SPREADSHEET ANALYSIS OF WALL STRESSES

PROBLEM: A railway track is to be built above an existing wall. The wall varies in height from 2.3 m to 5.1 m in height.

Due to space limitations the track is to be positioned with the nearest line of the track at 2.1 m from the top of the wall. The track is 1.5 m wide. The design loading from the train is 55 kN on each line. This is illustrated below.

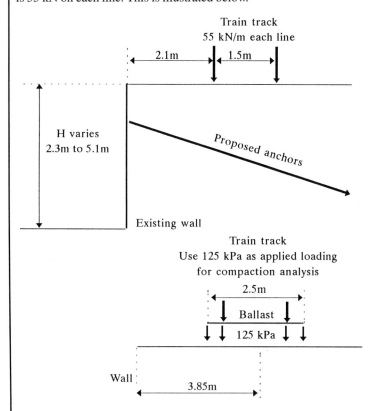

Boreholes reveal some loose to medium dense sand and gravel behind the wall (use an angle of friction of 30°).

Calculate the horizontal pressures experienced by the wall due to the line load and earth pressure effects. Given that the wall is structurally adequate, suitable wall anchors are required for this wall. The total thrust for the varying wall heights and the pressure distribution are required for design of the wall anchors.

In addition, the effect of the vibration of the train on the loose sand and gravel should be considered. (To allow for the effect of ballast under the lines, the load may be taken as 125 kPa on a 2.5 m wide track). Calculate the compaction induced pressures on the wall.

SOLUTION WORKSHEET: *Worksheets 4.1 and 4.2*

This shows the already completed worksheets from elastic analysis and compaction effects with earth pressure theory respectively. Use this as a guide during the development.

Some of the Lotus functions to be used are:
- ABSOLUTE ($) and relative addressing with Copy command.
- MATHEMATICAL functions:
  + X^2 is used for the square of X,
  + @SQRT(X) for the square root of X,
  + @PI for the mathematical constant,
- LOGICAL functions:
  + @IF (conditions compared, returns the value A if true, returns the value B if false). The IF is used together with the @ to activate the functional command. The value may be a number, text or formula.
- COMBINED analysis. Combining different methods of analysis in the same spreadsheet,
- GRAPHICAL analysis. Name and Use for multiple graphs in the same spreadsheet.

The first two were covered in Chapter 3 and will not be explained further in this chapter. Other features already explained in Chapter 3 will be expanded in this worksheet development.

This problem would ordinarily involve multiple calculations to cover the wide spectrum of possibilities, i.e. forces from elastic considerations, earth pressures and compaction effects. However with the ease of development of a spreadsheet then the full range of alternatives may be examined under one umbrella.

4.5.1 *General conditions*

The data input parameters must first be defined for the 3 various loading conditions.

General
- Job description,
- Wall height,
- Start of analysis depth + increment.

1. *Line load effect*
   - Line load(s),
   - Distance from wall.

2. *Earth pressure effects*
   - Active earth pressure coefficient,
   - Unit weight of backfill.

3. *Compaction induced effect*
   - Active earth pressure coefficient,

- Unit weight of backfill,
- Length of roller,
- Width of roller,
- Weight of roller,
- Centrifugal force,
- Distance of roller from wall.

The parameters (unit weight and active coefficient) listed in earth pressure theory are also found in the compaction induced effects, therefore these two conditions will be combined into the same worksheet. Two worksheets will be developed within the same spreadsheet:

Worksheet 1: Line load effects;

Worksheet 2: Compaction and earth pressure effects.

Table 4.8. Worksheet 4 layout.

Spreadsheet File name : WALL	
Worksheet 4.1 (Columns A to E) – Elastic theory	Worksheet 4.2 (Columns H to Q) – Compaction Theory
Data entry Form Computed Output Table Graphsheet – ELASTICLINE	Data Entry Form Computed Output Table Graphsheet – COMPACT
Graphsheet – COMBINED	

The data entry input is highlighted in the worksheet forms by use of the Range Unprotect commands for the data entry and at completion of the worksheet the Global Protection is used for the rest of the entries, whether text or formulae. The two worksheets should have an identification to separate the different types of analysis and a short explanation to identify its application. This is provided at the start of the worksheet. The development of the worksheets involves 3 steps for each of the loading conditions:

1. A data entry form,
2. A calculated output table, and
3. A graphsheet to provide the resulting graph.

Worksheet 4.1 is set up in columns A to E and shows the elastic stresses behind the wall. Worksheet 4.2 is set up from column H to Q and shows the compaction induced stresses behind the wall.

As both worksheets are developed on the same spreadsheet then the combined effect of all the various loading conditions may be illustrated on one graph for comparison purposes. In this example, the spreadsheet's file name is given as 'Wall' and the graphs shall be named 'Elasticline', 'Compact' and 'Combined'.

Use the Copy command in the replication of the formulae in the output table. Only the formula entry description is provided in full since the user is expected to be able to create the form and enter text from this point on.

The following decimal formats have been used in the calculated results to provide an easy to read worksheet screen:
   – Distances (metres): 1 decimal place,
   – Ratios: 3 decimal places,
   – Forces (kN): 1 decimal place,
   – Pressures (kPa): 2 decimal places.

Lotus command: / R Format Fixed {no of decimal places}

The two worksheets will now be explained separately.

   Worksheet 1 is shown as 4.1a and 4.1b – two screen displays – but this represents one continuous worksheet.

WORKSHEET 4.1: LATERAL EARTH PRESSURE DUE TO LINE LOADING

	A	B	C	D	E	F
1	Lateral Earth Pressure due to line loading					
2	--------	--------	--------	--------	--------	
3	This program calculates the horizontal pressure					
4	on rigid walls from a surface line load, using					
5	the modified Boussinesq equation.					
6						
7	Maximum Wall Height (H) =			5.1	metres	
8						
9	--------	--------	--------	--------	--------	
10			Q / unit length			
11	Line Loading		Load 1	Load 2	Combined	
12	--------	--------	--------	--------	++++++++	
13	Magnitude kN		55	55	110	
14	Distance x(metres)		2.1	3.6	********	
15	Ratio m = x/H		0.412	0.706	********	
16	Total thrust (kN)		29.9	23.4	53.3	
17	P(h) max (kPa)		5.87	4.58	10.45	
18	--------	--------	--------	--------	++++++++	
19	P(h) max = Equivalent Uniform Lateral Pressure					
20						

Worksheet 4.1a. Display screen

Worksheet 4.1 will be developed in the following steps:
   1. Create a worksheet description at the start.
   2. Create a data entry form for the design parameters mentioned previously.
   3. Provide intermediate calculations:
      – Ratio of offset distance to wall height;
      – Compute total thrust on wall;
      – Compute equivalent uniform lateral pressure on wall.
   4. Develop the output tables showing:
      – Table headings;
      – Depth variation below top of wall;
      – Ratio of depth to wall height;

	A	B	C	D	E	F
20						
21	+++++++++	+++++++++	+++++++++	+++++++++	+++++++++	
22	Depth	Ratio	Horizontal pressure (kPa)			
23	z (m)	n = z/H	Load 1	Load 2	Combined	
24	+++++++++	+++++++++	+++++++++	+++++++++	+++++++++	
25	0.0	0.000	0.00	0.00	0.00	
26	0.5	0.098	7.11	2.60	9.71	
27	1.0	0.196	10.55	4.66	15.21	
28	1.5	0.294	10.44	5.88	16.33	
29	2.0	0.392	8.73	6.31	15.04	
30	2.5	0.490	6.79	6.15	12.94	
31	3.0	0.588	5.15	5.65	10.80	
32	3.5	0.686	3.89	5.00	8.89	
33	4.0	0.784	2.97	4.33	7.29	
34	4.5	0.882	2.29	3.70	5.99	
35	5.0	0.980	1.79	3.15	4.93	
36	5.5	1.078	1.41	2.67	4.09	
37	6.0	1.176	1.13	2.27	3.41	
38	6.5	1.275	0.92	1.94	2.86	

Worksheet 4.1b. Display screen.

- Compute the horizontal pressures for the two line loads;
- Sum the pressures from the two conditions.
5. Convert the tabulated result to a graph 'Elasticline'.
6. Name this graph and save file 'Wall'.

The first step in the development after the worksheet description is the set up of the data entry form.

Steps 1 and 2: Data entry input form

The wall height, loads and offset distances are defined here.

A	B	C	D	E
Lateral Earth Pressure due to line loading				
---------	---------	---------	---------	---------
This program calculates the horizontal pressure				
on rigid walls from a surface line load, using				
the modified Boussinesq equation.				
Maximum Wall Height (H) =			5.1	metres
---------	---------	---------	---------	---------
		Q / unit length		
Line Loading		Load 1	Load 2	Combined
---------	---------	---------	---------	+++++++++
Magnitude kN		55	55	110
Distance x(metres)		2.1	3.6	*********

Worksheet 4.1. Data entry input form.

Step 3: Calculations

The displayed cells are sample output values. The actual cell entry values are explained below.

The ratio m (offset distance of load to wall height) is first calculated as this forms the defining condition upon which the relevant formulae as given in Tables 4.1 and 4.2 are used.

A	B	C	D	E
	Ratio m = x/H	0.412	0.706	*********
Total thrust (kN)		29.9	23.4	53.3
P(h) max (kPa)		5.87	4.58	10.45
---------	---------	---------	---------	+++++++++
P(h) max = Equivalent Uniform Lateral Pressure				

Worksheet 4.1. Intermediate calculations.

Unless this is defined initially, subsequent computations of the output table cannot proceed.

The ratio m in cell C15 can be entered in two ways:

(1) Type the entry: + C14 / D7, or

(2) Type +: Move the pointer to cell C14 (x distance)

Type /: Move the pointer to cell D7 (wall height).

The use of the + sign initially should be noted.

Cell (Format)	Display	Cell Entry Formula	Heading/Formula
C15 (F3)	0.412	+C14/D7	Ratio m (Load 1) / x/H
D15 (F3)	0.706	+D14/D7	Ratio m (Load 2) / x/H

Worksheet 4.1. Ratio *m*.

Cell D15 is similarly input. The range is then formatted to 3 decimal places.

The formulae of Table 4.2 calculate the total thrust on the wall. At cells C16 and D16 the value m is being checked for which of the two conditions the ratio m applies, which then invokes the relevant formula to calculate the thrust on the wall.

The two conditions are:

$m > 0.4$ and calculates: $2 Q_L / \pi / (m^2 + 1)$

$m \leq 0.4$ and calculates $0.548 Q_L$

The logical IF function is used to distinguish between the two conditions. The syntax of the formula is:

@IF (condition, Formula 1, Formula 2)

@IF (evaluates ratio $m > 0.4$, calculates: $2Q_L/\pi/(m^2 + 1)$, otherwise calculates: $0.548\,Q_L$).

This must be done for line loads 1 and 2. The result is then summed in cell E16.

Cell (Format)	Display	Cell Entry Formula	Heading/ Formula
C16 (F1)	29.9	@IF(C15>0.4, 2*C13/@PI/(C15^2+1),0.548*C13)	Total thrust (load 1) / REF : Table 4.2
D16 (F1)	23.4	@IF(D15>0.4, 2*D13/@PI/(D15^2+1),0.548*D13)	Total thrust (load 1) / REF : Table 4.2
E16 (F1)	53.3	+C16+D16	Thrust1 + Thrust2

Worksheet 4.1. Total thrust.

This spreadsheet has been developed to assess the effect of a train track (i.e. 2 line loads) passing adjacent to an existing wall. For a problem with a single line load only, one of the two loads may be input as zero, irrespective of distance. Alternatively, the distance may be placed to infinity irrespective of the load.

The equivalent uniform lateral pressure is calculated by dividing the total thrust on the wall by the given wall height (H). Cell E17 is the summation of the individual pressures produced by the two line loads.

The procedure adopted to this point would typically be used in calculations for a specified wall height. However, in this problem there is a varying wall height, with the 5.1 metres representing *only the maximum condition*.

The problem requires that an existing wall of significant length use tie back anchors to take an additional loading condition. For the entire length designed on the extreme 5.1 metre height then it would be an overdesign, which satisfies simplicity in calculations and/or construction. A detailed analysis should cover the range of wall heights to maintain economies in design.

The next step is then to produce a table of conditions for various depths of wall and this also serves to produce the shape of the pressure distribution diagram.

Step 4: Output table

The depth axis is set up using 0.5 metre increments. The actual depths may be varied by changing the contents of the first cell A25. The other cells would change accordingly.

Alternatively the start depth and increment may have been defined at the start

A	B	C	D	E
+++++++++	+++++++++	+++++++++	+++++++++	+++++++++
Depth	Ratio	Horizontal pressure (kPa)		
z (m)	n = z/H	Load 1	Load 2	Combined
+++++++++	+++++++++	+++++++++	+++++++++	+++++++++
0.0	0.000	0.00	0.00	0.00
0.5	0.098	7.11	2.60	9.71
1.0	0.196	10.55	4.66	15.21
1.5	0.294	10.44	5.88	16.33
2.0	0.392	8.73	6.31	15.04
2.5	0.490	6.79	6.15	12.94
3.0	0.588	5.15	5.65	10.80
3.5	0.686	3.89	5.00	8.89
4.0	0.784	2.97	4.33	7.29
4.5	0.882	2.29	3.70	5.99
5.0	0.980	1.79	3.15	4.93
5.5	1.078	1.41	2.67	4.09
6.0	1.176	1.13	2.27	3.41
6.5	1.275	0.92	1.94	2.86

Worksheet 4.1. Tabulated output.

Cell (Format)	Display	Cell Entry Formula	Heading/ Formula
A25 (F1)	0	0	
A26 (F1)	0.5	+A25+0.5	
A27 (F1)	1.0	+A26+0.5	z + 0.5
A28 (F1)	1.5	+A27+0.5	
etc.			

Worksheet 4.1. Depth z.

of the worksheet as in Chapter 3. This would produce a more flexible spreadsheet in being able to define the start of the analysis as well as the increment for analysis. However, in this instance, the 'typical' values given (0.0 metre depth and 0.5 metre increments) would seldom be varied.

The depths shown do not relate directly to the actual height of the wall, but the pressure distribution for an arbitrary height. This helps to provide the shape of the distribution curve and acts as a check on the sensitivity of the model to the effects of varying wall heights.

Note that the format of this range has been set to 1 decimal place.

Column B calculates the n (= z/H) ratio that is subsequently used in the formulae of the adjoining columns. At $n = 1.0$ the full height of the wall is realised.

Cell (Format)	Display	Cell Entry Formula	Heading/ Formula
B25 (F3)	0.000	+A25/$D$7	
B26 (F3)	0.098	+A26/$D$7	
B27 (F3)	0.196	+A27/$D$7	z/H
B28 (F3)	0.294	+A28/$D$7	
etc.			

Worksheet 4.1. Ratio $n = z/H$.

Columns C and D calculate the effect of line loads 1 and 2 respectively while column E is the effect of both loads. The ratio m (offset distance of load to wall height) is being checked to apply the relevant formula of Table 4.1 to calculate the horizontal stresses.

Cell (Format)	Display	Cell Entry Formula	Heading/ Formula
C25 (F2)	0.00	@IF($C$15>0.4,4*$C$13/@PI/$D$7* ($C$15*$C$15*B25/($C$15^2+B25^2)^2),$C$13/$ D$7*0.203*B25/(0.16+B25^2)^2)	REF : Table 4.1
C26 (F2)	7.11	@IF($C$15>0.4,4*$C$13/@PI/$D$7* ($C$15*$C$15*B26/($C$15^2+B26^2)^2),$C$13/$ D$7*0.203*B26/(0.16+B26^2)^2)	
C27 (F2)	10.55	@IF($C$15>0.4,4*$C$13/@PI/$D$7* ($C$15*$C$15*B27/($C$15^2+B27^2)^2),$C$13/$ D$7*0.203*B27/(0.16+B27^2)^2)	
etc.		etc.	
D25 (F2)	0.00	@IF($D$15>0.4,4*$D$13/@PI/$D$7* ($D$15*$D$15*B25/($D$15^2+B25^2)^2),$D$13/$ D$7*0.203*B25/(0.16+B25^2)^2)	REF : Table 4.1
D26 (F2)	2.60	@IF($D$15>0.4,4*$D$13/@PI/$D$7* ($D$15*$D$15*B26/($D$15^2+B26^2)^2),$D$13/$ D$7*0.203*B26/(0.16+B26^2)^2)	
D27 (F2)	4.66	@IF($D$15>0.4,4*$D$13/@PI/$D$7* ($D$15*$D$15*B27/($D$15^2+B27^2)^2),$D$13/$ D$7*0.203*B27/(0.16+B27^2)^2)	
etc.		etc.	
E25 (F2)	0.00	+C25+D25	Sum of the
E26 (F2)	9.71	+C26+D26	two
E27 (F2)	15.21	+C27+D27	pressures
etc.		etc.	

Worksheet 4.1. Horizontal pressure.

This involves the use of the @IF command. Cell C25 only is typed, then copied to the columns below.

In the regular mathematical approach this is the input of the following parameters:

IF $m > 0.4$ $\qquad\qquad (4\,Q_L/\pi H) * m^2 n/\,(m^2 + n^2)^2$

ELSE (IF $m < = 0.4$) $\qquad (0.203\,Q_L/H) * n/(0.16 + n^2)^2$

The applied loading Q_L and the *m* value are used as absolute values. The *n* value is used as the variable in the cell input. When this cell value is copied along column C then the only adjustment which would occur is the depth factor (relative addressing) while the other 'variables' are kept constant (absolute addressing).

A similar logic is repeated in the input of the formula in columns D and E, but with the magnitude, distance and *m* value of load 2. Note that rather than repeating the lengthy formula input into cell D25, it is easier to copy cell C25 and then use the EDIT {F2} function.

Column E is the summation of the two line load effects. Column C value is added to column D to produce the net effect of the two rail lines of the track.

Step 5: Graphical output

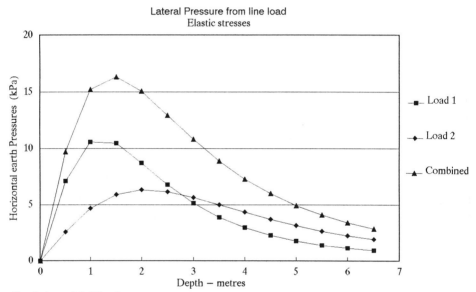

Graphsheet 4.1. Elastic pressures.

A graph can now be produced from the tabular values as shown in Graphsheet 4.1 using:

```
/ Graph
        X (A25.A38)                                    ...Depth
        A (C25.C38)                                   ...Load 1
        B (D25.D38)                                   ...Load 2
        C (E25.E38)                                ...Combined.
```

The graph should be named in order to have multiple graphs within the same spreadsheet and eliminate the need to constantly redo the defined graph parameters. This is accomplished by:

```
/ Graph Name Create {Filename – Elasticline}
```

If another graph had been current and the user wished to view the earlier graph, then use:

```
/ Graph Name Use {Filename – Elasticline}
```

When the worksheet file is saved, multiple graphs may be saved as well, if it is named during the graph set up.

This completes the development of Worksheet 4.1. The complete listing is provided in the appendix.

WORKSHEET 4.2: LATERAL EARTH PRESSURE DUE TO COMPACTION EFFORT

This worksheet computes:
- the earth pressure effects, and
- the compaction induced pressures.

The granular backfill behind the wall was found to be loose soil in some places and the effect of the train vibrations may produce compaction induced pressures. This worksheet calculates these pressures.

The worksheets shown are broken up for the purposes of this text, and it is emphasised that the entire worksheet 4.2a, 4.2b, 4.2c and 4.2d is continuous.

NOTE: The entire worksheet may be viewed in one screen by modifying the 123.set file, before initialisation of the Lotus program. The high resolution driver set (say high-set) may then be activated by the command "Lotus high" at the DOS prompt. The larger view of the worksheet screen is also available with add-on packages.

Lotus 3.X allows a wider view of the screen, as well as providing the flexibility of placing the second analysis on the worksheet behind the first.

WYSIWYG Add-in allows control of the number of rows displayed on a screen while in graphics mode. Use: `Display Rows` {Type a number between 16 to 60}.

NOTE: Lotus 3.X users would have the option here of using a separate sheet to create this worksheet within the same file. This would be done by using: `\ Worksheet Insert Sheet (Before or After)`.

Worksheet design

The compaction induced stresses are set up in the same manner as the line load pressures in Worksheet 4.1. However, this worksheet starts at column I. Again the @IF logic function is used to activate the relevant formula, but using Table 4.6.

Figure 4.3 showed that the magnitude of the horizontal pressure is dependent on the depth and with 3 distinct variations of the pressure distribution as it meets

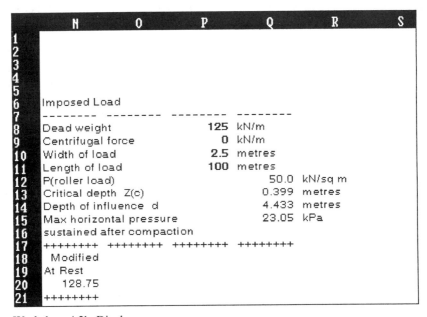

	H	I	J	K	L	M
1		Lateral Earth Pressure due to compaction effort				
2		-------- -------- -------- -------- --------------				
3		This program calculates the horizontal pressure on rigid walls				
4		from a surface load, with compaction induced effects				
5						
6		Geometry & Soil Properties				
7		-------- -------- --------				
8		Wall Height (H)		5.1	metres	
9		Distance from wall		3.85	metres	
10						
11		Unit weight (kN/cu m)			18	
12		Compact unit wght (kN/cu m)			19	
13		Minimum unit wght (kN/cu m)			17	
14		Active earth pressure Ka			0.3	
15		At Rest earth pressure Ko			0.5	
16		Compacted at rest press Koc			0.55	
17		++++++++ ++++++++ ++++++++ ++++++++ +++++++++++++				
18						Compaction
19			Active	At Rest		Induced
20		Total thrust (kN)	70.23	117.04		130.13
21		++++++++ ++++++++ ++++++++ ++++++++ +++++++++++++				

Worksheet 4.2a. Display screen.

	N	O	P	Q	R	S
1						
2						
3						
4						
5						
6	Imposed Load					
7	-------- -------- -------- --------					
8	Dead weight		125	kN/m		
9	Centrifugal force		0	kN/m		
10	Width of load		2.5	metres		
11	Length of load		100	metres		
12	P(roller load)			50.0	kN/sq m	
13	Critical depth Z(c)			0.399	metres	
14	Depth of influence d			4.433	metres	
15	Max horizontal pressure			23.05	kPa	
16	sustained after compaction					
17	++++++++ ++++++++ ++++++++ ++++++++					
18	Modified					
19	At Rest					
20	128.75					
21	++++++++					

Worksheet 4.2b. Display screen.

	I	J	K	L	M	N
18					Compaction	Modified
19			Active	At Rest	Induced	At Rest
20	Total thrust (kN)		70.23	117.04	130.13	128.75
21	+++++++++	++++++++	+++++++++	+++++++++	++++++++++++++	++++++++
22			Horizontal pressure (kPa)			
23	Depth z (metres)		Active	At Rest	Induced		Modified
24	+++++++++	++++++++	+++++++++	+++++++++	++++++++++++++	++++++++
25	0.0		0.00	0.00	0.00	0.00
26	0.5		2.70	4.50	23.05	4.95
27	1.0		5.40	9.00	23.05	9.90
28	1.5		8.10	13.50	23.05	14.85
29	2.0		10.80	18.00	23.05	19.80
30	2.5		13.50	22.50	23.05	24.75
31	3.0		16.20	27.00	23.05	29.70
32	3.5		18.90	31.50	23.05	34.65
33	4.0		21.60	36.00	23.05	39.60
34	4.5		24.30	40.50	24.30	44.55
35	5.0		27.00	45.00	27.00	49.50
36	5.5		29.70	49.50	29.70	54.45
37	6.0		32.40	54.00	32.40	59.40
38	6.5		35.10	58.50	35.10	64.35

Worksheet 4.2c. Display screen.

	I	J	K	L	M	N
39						
40						
41						
42	Height of		Total Thrust (kN)			
43	Wall (m)		Active	At Rest	Induced		Modified
44	+++++++++	++++++++	+++++++++	+++++++++	++++++++++++++	++++++++
45	0		0.00	0.00	0.00	0.00
46	1.0		2.70	4.50	18.45	4.95
47	2.0		10.80	18.00	41.50	19.80
48	3.0		24.30	40.50	64.55	44.55
49	4.0		43.20	72.00	87.60	79.20
50	5.0		67.50	112.50	125.10	123.75
51	6.0		97.20	162.00	177.85	178.20
52						
53						
54						
55						
56						
57						
58						

Worksheet 4.2d. Display screen.

the active earth pressure (K_a) condition with depth. The @IF function is used to differentiate between the 3 conditions.

Note the display values as against the cell entry formula. If the formula is to be displayed instead of the values use:

```
/ Range Format Text
```

The specified range will then display the actual formula entered. Unless the column width settings are changed the entire formula may not be fully displayed. The cell values may be displayed again by:

```
/ Range Format Reset
```

This may upset the Range Format Fixed settings, when the number of decimal places were specified and would require to be reset.

Data entry
The data entry involves defining the geometry, soil properties and the imposed loads.

This is seen in rows 1 to 17.

I	J	K	L	M
Lateral Earth Pressure due to compaction effort				
---------	---------	---------	---------	-------------
This program calculates the horizontal pressure on rigid walls				
from a surface load, with compaction induced effects				
Geometry & Soil Properties				
---------	---------	---------		
Wall Height (H)		5.1	metres	
Distance from wall		3.85	metres	
Unit weight (kN/cu m)			18	
Compact unit wght (kN/cu m)			19	
Minimum unit wght (kN/cu m)			17	
Active earth pressure Ka			0.3	
At Rest earth pressure Ko			0.5	
Compacted at rest press Koc			0.55	

Worksheet 4.2. Rows 1 to 17: Geometry and soil properties.

In order to maintain the flow of the worksheet development, intermittent calculations are made. These results will then be used in the generation of the results table.

Calculated result

The maximum compacted horizontal pressures and the compacted at rest pressure coefficient are calculated using the given formulae (modified K_o approach) and entered at cell L16.

N	O	P	Q	R
Imposed Load				
---------	----------	----------	----------	
Dead weight			125 kN/m	
Centrifugal force			0 kN/m	
Width of load			2.5 metres	
Length of load			100 metres	
P(roller load)				50.0 kN/sq m
Critical depth Z(c)				0.399 metres
Depth of influence d				4.433 metres
Max horizontal pressure				23.05 kPa
sustained after compaction				
+++++++++	+++++++++	+++++++++	+++++++++	

Worksheet 4.2. Rows 7 to 18: Imposed loads.

Cell (Format)	Display	Cell Entry Formula	Heading/ Formula
L16	0.55	+L15+(L12−L13)*0.025	Table 4.5

Worksheet 4.2. Compacted at rest pressure koc.

The entry and calculations for the imposed loading then follows in rows 7 to 18 and columns N, O, P, Q, and R (Table 4.7).

Critical depth $= K_a \sqrt{(2*P/\pi/\gamma)}$

The critical depth and depth of influence are determined from the depths defined in Table 4.7 and illustrated in Figure 4.3.

This will act as the change points for the application of the different formulae to calculate the resulting pressures and forces.

Use of the Lotus commands: @SQRT and $ (ABSOLUTE) should be noted.

Cell (Format)	Display	Cell Entry Formula	Heading/ Formula
Q12	50.0	(P8+P9)/P10	P (Roller Load)
Q13	0.399	+$L$14*@SQRT(2*$Q$12/@PI/$L$11)	Critical Depth
Q14	4.433	@SQRT(2*$Q$12/@PI/$L$11)/$L$14	Depth of Influence
Q15	23.05	@SQRT(2*Q12*L11/@PI)*P11/(K9+P11)	Max horizontal pressure

Worksheet 4.2. Calculation of change point depths and max. pressures.

Computed results

The input data is now used to produce a table of values for a range of wall heights.

The total thrust represents the areas under the respective pressure diagrams for the specified wall height, and is calculated here for the maximum wall height of 5.1 metres.

I	J	K	L	M	N
				Compaction	Modified
		Active	At Rest	Induced	At Rest
Total thrust (kN)		70.23	117.05	130.13	128.75

Worksheet 4.2. Rows 18 to 20: Display screen.

Cell (Format)	Display	Cell Entry Formula	Heading/ Formula
K20	70.23	+$L$14*$L$11*$K$8^2/2	Active
L20	117.05	+$L$15*$L$11*$K$8^2/2	At Rest
M20	130.13	@IF($K$8<$Q$13, $K$8/$Q$13*$Q$15*$K$8/2, @IF($K$8<$Q$14,$Q$15*($K$8-$Q$13)+$Q$15* $Q$13/2,$Q$15*($K$8-$Q$13)+($Q$15*$Q$13/2) +($K$8-$Q$14)*$L$14*$L$11*($K$8+$Q$14)/2))	Compaction Induced
N20	128.75	+$L$16*$L$11*$K$8^2/2	Modified At rest

Worksheet 4.2. Rows 18 to 20: Formula entry.

The 4 different possibilities for analysis – active, at rest, compaction induced, modified at rest are examined. Only the explanation of the formula in cell M20 is given.

The use of multiple embedded @IF commands is required because of the 3 stages of the compaction induced pressures:

$$z < Z_c$$
$$Z_c < z < d$$
$$z > d$$

As the compaction induced pressure diagram varies with depth, the height of the wall is compared with the critical depth (z_c) and the depth of influence (d) to determine which area of the pressure diagram is applicable to compute the total thrust.

These depths had been previously determined (Q13 and Q14). The logic function using the Lotus format is:

@IF ($H < z_c$, Formula 1, @IF ($H < d$, Formula 2, Formula 3)

NOTE: IF wall height (H).	Using formula given in Table 4.7 (Fig. 4.4).
1. $H <$ critical depth z_c	$(z/z_c) *$ top pressure
2. $H <$ influence depth d	$z_c *$ top pressure +
	$(z-z_c) *$ Max sustained after compaction
3. Otherwise $(H > d)$	$z_c *$ top pressure +
	$(d-z_c) *$ max sustained after compaction +
	$(z-d) *$ active pressures

The calculations to this point would be typical for a specified wall height in the routine design work. This would then be repeated for varying wall heights or alternatively the entire wall design for the maximum wall height condition for simplicity of construction.

The tables which follow in rows 25 to 38 show the wall pressures with *varying* wall heights. This is the main output table for this analysis.

The variation of the pressures with depth is tabulated to an arbitrarily chosen depth and is evaluated. (This depth 6.5 metres may have been defined as a variable

Cell (Format)	Display	Cell Entry Formula	Heading/ Formula
I25	0.0	0.00001	Depth
I26	0.5	+I25+0.5	
I27	1.0	+I26+0.5	
I28	1.5	+I27+0.5	
etc.		etc.	
K25	0.00	+$L$14*$L$11*I25	Active Pressures
K26	2.70	+$L$14*$L$11*I26	
K27	5.40	+$L$14*$L$11*I27	
K28	8.10	+$L$14*$L$11*I28	
etc.		etc.	
L25	0.00	+$L$15*$L$11*I25	Modified At Rest
L26	4.50	+$L$15*$L$11*I26	
L27	9.00	+$L$15*$L$11*I27	
L28	13.50	+$L$15*$L$11*I28	
etc.		etc.	
M25	0.00	@IF(I25<$Q$13,I25/$Q$13*$Q$15, @IF(I25<$Q$14,$Q$15,K25))	Induced Compaction
M26	23.05	@IF(I26<$Q$13,I26/$Q$13*$Q$15, @IF(I26<$Q$14,$Q$15,K26))	
M27	23.05	@IF(I27<$Q$13,I27/$Q$13*$Q$15, @IF(I27<$Q$14,$Q$15,K27))	
M28	23.05	@IF(I28<$Q$13,I28/$Q$13*$Q$15, @IF(I28<$Q$14,$Q$15,K28))	
etc.		etc.	
N25	0.00	+$L$16*$L$11*I25	At Rest Pressures
N26	4.95	+$L$16*$L$11*I26	
N27	9.90	+$L$16*$L$11*I27	
N28	14.35	+$L$16*$L$11*I28	
etc.		etc.	

Worksheet 4.2. Rows 25 to 38: Formula entry.

Cell (Format)	Display	Cell Entry Formula	Heading/ Formula
K45	0.00	$L$14*$L$11*I45^2/2	Active
K46	2.70	+$L$14*$L$11*I46^2/2	
K47	10.8	+$L$14*$L$11*I47^2/2	
etc		etc.	
L45	0.00	+$L$15*$L$11*I45^2/2	At Rest
L46	4.50	+$L$15*$L$11*I46^2/2	
L47	18.0	+$L$15*$L$11*I47^2/2	
etc		etc.	
M45	0.00	@IF(I45<$Q$13,I45/$Q$13*$Q$15*I45/2, @IF(I45<$Q$14,$Q$15*(I45−$Q$13)+$Q$15*$Q$1 3/2,$Q$15*(I45−$Q$13)+($Q$15*$Q$13/2)+(I45−$ Q$14)*$L$14*$L$11*(I45+$Q$14)/2))	Induced
M46	18.4	@IF(I46<$Q$13,I46/$Q$13*$Q$15*I46/2, @IF(I46<$Q$14,$Q$15*(I46−$Q$13)+$Q$15*$Q$1 3/2,$Q$15*(I46−$Q$13)+($Q$15*$Q$13/2)+(I46−$ Q$14)*$L$14*$L$11*(I46+$Q$14)/2))	
M47	41.5	@IF(I47<$Q$13,I47/$Q$13*$Q$15*I47/2, @IF(I47<$Q$14,$Q$15*(I47−$Q$13)+$Q$15*$Q$1 3/2,$Q$15*(I47−$Q$13)+($Q$15*$Q$13/2)+(I47−$ Q$14)*$L$14*$L$11*(I47+$Q$14)/2))	
etc.		etc.	
N45	0.00	+$L$16*$L$11*I45^2/2	Modified
N46	4.95	+$L$16*$L$11*I46^2/2	
N47	9.90	+$L$16*$L$11*I47^2/2	
etc.		etc.	

Worksheet 4.2. Rows 45 to 51: Formula entry.

if required – but was chosen simply because most walls would fall within that range).

The actual formula entry for the output table is shown.

The active and at rest pressures are calculated from the formula of Table 4.4 but with the respective pressure coefficients.

The modified at rest pressures are calculated from the formula of Table 4.5.

The induced compaction pressures are calculated from the formula of Table 4.6 and allowing for change point depths (Table 4.7).

The total thrust for wall height of 5.1 metres was calculated at row 20, and explained previously. As an addendum to the spreadsheet, Worksheet 4.2c shows the calculated *total thrusts* for varying depths. It is merely an expansion for parametric evaluation of the varying wall height.

The explanation of the above formulae follows from the computed results explanation of the total thrust calculated at rows 18 to 20.

Graphsheet
The graphsheet is then generated from the calculated table of I25.N38.

```
/ Graph
        X (I25.I38)                                    ...Depth
        A (K25.K38)                                    ...Active
        B (L25.L38)                                    ...At rest
        C (M25.M38)                          ...Induced compaction
        D (N25.N38)                             ...Modified at rest.
```

The graph should be named.

```
/ Graph Name {Filename – Compact}
```

The title of the graph is then set and shown in Graphsheet 4.2 for the variation of horizontal pressures with depth of wall.

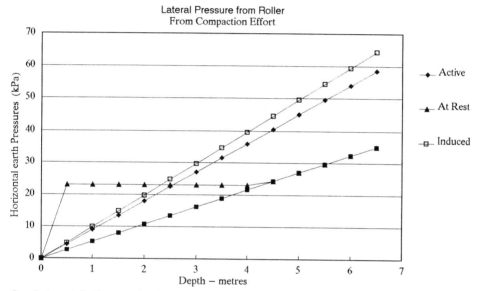

Graphsheet 4.2. Compaction induced pressures.

Combined results
The results of all the calculated lateral pressures may be combined into one graph by changing only the variable graph references.

```
/ Graph
        X (A25.A38)                                    ...Depth
        A (K25.K38)                                    ...Active
        B (L25.L38)                                    ...At rest
        C (M25.M38)                          ...Induced compaction
        D (N25.N38)                             ...Modified at rest
        E (E25.E38)                             ...Elastic stresses.
```

The combined effect of the elastic and compaction induced effects are presented in Graphsheet 4.3.

The graph should be named to distinguish between the other two named graphs of the spreadsheet.

/ Graph Name {Filename – Combined}

The spreadsheet has now been set up so that any future changes can be evaluated by keying in the new entry. By using the {F10} function key then the new graph may be observed.

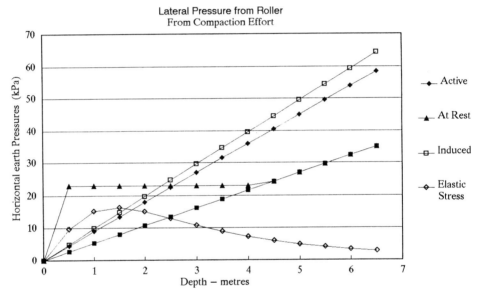

Graphsheet 4.3. Combined pressures.

For analysis, only the filling of the worksheet form is required to produce the tabulated result and graphs.

Save the file before exiting the spreadsheet:

/ File Save Wall

The analysis of the problem given may now proceed with a full spread of results for varying height walls as well as different methods of analysis.

```
Lateral Earth Pressure due to line loading
-------------------------------------------------------

This program calculates the horizontal pressure
on rigid walls from a surface line load, using
the modified Boussinesq equation.

Maximum Wall Height (H) =                    5.1 metres

-------------------------------------------------------

                       Q / unit length
Line Loading           Load 1     Load 2       Combined
--------------------------------------------------++++++++·
Magnitude kN               55         55          110
Distance x(metres)        2.1        3.6 *************
Ratio m = x/H           0.412      0.706 *************
Total thrust (kN)        29.9       23.4          53.3
P(h) max (kPa)           5.87       4.58         10.45
----------------------------------------------++++++++·
P(h) max = Equivalent Uniform Lateral Pressure

++++++++·++++++++·++++++++·++++++++·++++++++·
  Depth      Ratio    Horizontal pressure (kPa)
  z (m)     n = z/H      Load 1    Load 2      Combined
++++++++·++++++++·++++++++·++++++++·++++++++·
   0.0      0.000        0.00      0.00        0.00
   0.5      0.098        7.11      2.60        9.71
   1.0      0.196       10.55      4.66       15.21
   1.5      0.294       10.44      5.88       16.33
   2.0      0.392        8.73      6.31       15.04
   2.5      0.490        6.79      6.15       12.94
   3.0      0.588        5.15      5.65       10.80
   3.5      0.686        3.89      5.00        8.89
   4.0      0.784        2.97      4.33        7.29
   4.5      0.882        2.29      3.70        5.99
   5.0      0.980        1.79      3.15        4.93
   5.5      1.078        1.41      2.67        4.09
   6.0      1.176        1.13      2.27        3.41
   6.5      1.275        0.92      1.94        2.86
```

Worksheet form 4.1. Data entry and output table. Filename: Wall.

Lateral Earth Pressure due to compaction effort

This program calculates the horizontal pressure on rigid walls
from a surface load, with compaction induced effects

Geometry & Soil Properties			Imposed Load	
Wall Height (H)	5.1	metres	Dead weight	125 kN/m
Distance from wall	3.85	metres	Centrifugal force	0 kN/m
			Width of load	2.5 metres
Unit weight (kN/cu m)	18		Length of load	100 metres
Compact unit wght (kN/cu m)	19		P(roller load)	50.0 kN/sq m
Minimum unit wght (kN/cu m)	17		Critical depth $Z(c)$	0.399 metres
Active earth pressure Ka	0.3		Depth of influence d	4.433 metres
At Rest earth pressure Ko	0.5		Max horizontal pressure	23.05 kPa
Compacted at rest press Koc	0.55		sustained after compaction	

++++++ ++++++ +++++++ +++++++ +++++++++ ++++++ ++++++ +++++ +++++

	Active	At Rest	Compaction Induced	Modified At Rest
Total thrust (kN)	70.23	117.05	130.13	128.75

++++++ ++++++ +++++++ +++++++ +++++++++ ++++++

Horizontal pressure (kPa)

Depth z (metres)	Active	At Rest	Induced	Modified
0.0	0.00	0.00	0.00	0.00
0.5	2.70	4.50	23.05	4.95
1.0	5.40	9.00	23.05	9.90
1.5	8.10	13.50	23.05	14.85
2.0	10.80	18.00	23.05	19.80
2.5	13.50	22.50	23.05	24.75
3.0	16.20	27.00	23.05	29.70
3.5	18.90	31.50	23.05	34.65
4.0	21.60	36.00	23.05	39.60
4.5	24.30	40.50	24.30	44.55
5.0	27.00	45.00	27.00	49.50
5.5	29.70	49.50	29.70	54.45
6.0	32.40	54.00	32.40	59.40
6.5	35.10	58.50	35.10	64.35

Height of Wall (m)	Total Thrust (kN) Active	At Rest	Induced	Modified
0	0.00	0.00	0.00	0.00
1.0	2.70	4.50	18.45	4.95
2.0	10.80	18.00	41.50	19.80
3.0	24.30	40.50	64.55	44.55
4.0	43.20	72.00	87.60	79.20
5.0	67.50	112.50	125.10	123.75
6.0	97.20	162.00	177.85	178.20

Worksheet form 4.2. Data entry and output table. Filename: Wall.

CHAPTER 5

Slope stability

Slopes may be natural, formed by excavation or by the build up of construction materials in embankments and earth dams. Instability may be caused by gravitational, imposed or seepage forces and/or a change in material properties.

Planar analysis is a special case of a wedge analysis. A plane failure is often used in analysis because of the ease of calculations from simplicity in the geometry and loading input. The analysis of planar rock slopes will be covered in the spreadsheet development to show the sensitivity of the slope to some of the simplifying assumptions typically made in analysis.

Use of the graphical analysis integrated with the spreadsheet design is expanded in this chapter. Further uses of logic functions are shown. The graphical output of the analysis would result in a suitable slope design, since the range of 'what-if' would form part of the analysis. The sensitivity of the design to the various parameters is then seen.

5.1 THEORY OF SLOPES

Limiting equilibrium methods are used in the analysis of slope instability and involves evaluation of resisting forces with the forces tending to cause instability. The slope may have imposed loads such as surcharge, earthquake loading, as well as restraining loads such as reinforcement by the use of anchors or buttress support.

The factors affecting the stability are therefore:
1. The slope geometry,
2. Imposed loading and support,
3. The soil/rock properties,
4. The water conditions.

These will now be dealt with in turn for a slope of unit width. Reference should be made to Hoek & Bray (1981).

5.1.1 *Slope geometry*

The slope geometry is based on:
- the (exposed) slope face angle (ψ_f),
- the slope plane (failure) angle (ψ_p), (ψ_a), and (ψ_b) for wedges,
- tension cracks of depth (Z),
- location (D) of the tension crack (face or crown),
- height of slope (H).

This is defined in Figure 5.1. Calculated parameters would include the plane length (A), crown length (X) and location of tension crack (D).

The failure surface (slope plane angle) and block size would be limited by the presence of the tension crack. These tension cracks may be fully or partially filled with water, resulting in water pressures which add to the factors causing slope instability. The tension cracks would also limit the activating force of the block self weight.

The analysis assumes a flat and horizontal crown.

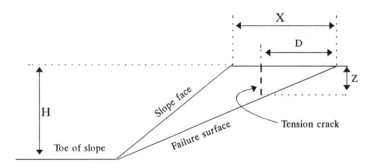

Slope with tension crack at crown

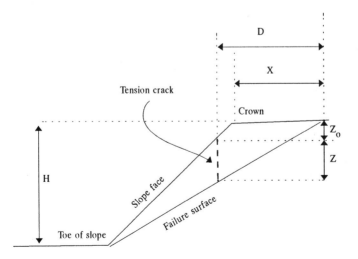

Slope with tension crack at face

Figure 5.1. Slope geometry.

Table 5.1. Weight of wedge.

Location of tension crack	Weight
Slope crown	$\frac{1}{2} * \gamma * [HX - DZ]$
Slope face	$\frac{1}{2} * \gamma * [HX - DZ + Z_o(D - X)]$

The weight of the wedge is given in Table 5.1 for the two possible cases of the tension crack in the crown or face of the slope.

5.1.2 *Imposed loading and support*

In addition to the natural gravitational force which is an activating condition and the natural bond (cohesion and friction) which would resist movement, applied loads may also act to restrain the slope or to activate it. These external forces are summarised in Table 5.2 with the support mechanisms illustrated in Figure 5.2.

The imposed loading condition may be:
 – a column at the top of an excavation (point load),
 – a wall along the crown (strip/line with a possible lateral loading transferred to the rock mass),
 – fill placed at the crown of the slope (uniform surcharge).
The supporting conditions may be:
 – rock anchors,
 – buttress to hold the wedge.
The lateral force (L) acts horizontally as an activating force, and may be the lateral force of a wall at the crown of the slope, earthquake load, etc. The buttress support (B) also acts horizontally but as a restraining force.

The buttress (Fig. 5.2a) at the toe of the slope provides a lateral restraint through dead weight. It also increases the strength of the material below the buttress by increasing the normal stress. An important requirement of buttress fill material is that it should be free draining.

Rock anchors (Fig. 5.2b) provide support partially by direct resistance to

Table 5.2. External loads.

Type of force	Condition
Surcharge q	Activating
Point load Q	
Lateral load L	
Anchor support T	Resisting
Buttress support B	

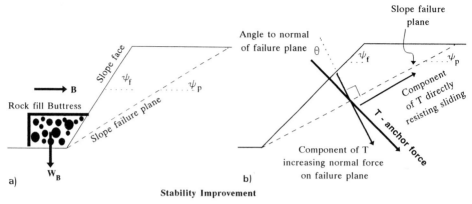

Figure 5.2a. Buttress support, b. Anchor support.

sliding (i.e. breaking force of the anchor strands for the applied shear force) and partly through the increased normal stress across discontinuities which provides a corresponding increase in frictional resistance to sliding.

The angle (α) of the anchor force to the normal of the slope face is given by $\alpha = \theta - (\psi_f - \psi_p)$.

5.1.3 *The soil/rock properties*

The soil/rock properties are determined from:
- cohesion along the plane of movement: c (kPa),
- the angle of friction: ϕ (degrees),
- the unit of weight of the rock mass: γ (kN/cu m).

5.1.4 *The water conditions*

The water conditions are assessed from the depth of water in the tension crack (Z_w) with four possible conditions:
- Dry slope: $Z_w = 0$ (an extreme condition),
- Water in the tension crack only,
- Water in the tension crack and on the sliding surface,
- Saturated slope with heavy recharge $Z_w = Z$ (an extreme condition).

Water forces are generated by the tension cracks as horizontal (V) water forces and an uplift (U) water force along the failure plane perpendicular to that plane. As water can exit from the face, then the pressure (U) is zero at the slope face. The forces may include any combination of the water + imposed forces listed in Table 5.2.

The four conditions of the water forces are summarised in Table 5.3 and illustrated in Figure 5.3.

Table 5.3. Water forces.

Water condition	Water force Horizontal (V)	Water force Uplift (U)
1. Dry slope	0	0
2. Tension crack only	$\frac{1}{2} * \gamma_w * Z_w^2$	0
3. Tension crack and on sliding surface	$\frac{1}{2} * \gamma_w * Z_w^2$	$\frac{1}{2} * \gamma_w * Z_w * A$
4. Saturated slope with heavy recharge	$\frac{1}{2} * \gamma_w * Z_w^2$	$\frac{1}{2} * \gamma_w * Z_w * A$

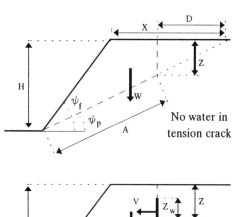

Dry slope

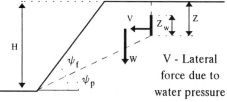

Water in tension crack only

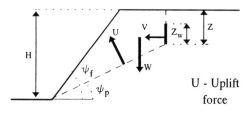

Water in tension crack & sliding surface

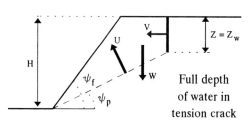

Saturated slope with heavy recharge

Figure 5.3. Water conditions.

Dry slope: $\Rightarrow V = 0;\ U = 0$
Water in a tension crack only: $\Rightarrow V$ from $z_w;\ U = 0$
Water in tension crack and sliding surface: $\Rightarrow V$ from $z_w;\ U$
Saturated and with heavy recharge: General condition:
                                                                   $\Rightarrow V\ (\text{max})$ from $z = z_w;\ U\ (\text{max})$

It is evident that a wide range of variables must be considered in the analysis. This lends itself to spreadsheet analysis. The slope conditions may be analysed to produce the factor of stability (instability) for any combination of the above variables, depending on which is the most sensitive or least known parameter.

5.2 SLOPE ANALYSIS

The analysis of slopes usually involves calculating the factor of safety of the slope. The factor of safety (F) is the ratio of the total force resisting sliding to the total force tending to induce the sliding as given in Table 5.4a.

For the condition of an external support with an anchor of magnitude T acting at an angle θ to the failure plane and/or a buttress toe support B then the limiting equilibrium equations are given in Table 5.4b.

This is the general equation which considers all of the mentioned forces from limit equilibrium. If a force does not apply (e.g. for no buttress support, $B = 0$) the equation reduces to that of Table 5.4a.

Table 5.4a. Natural forces only.

Resisting force $= cA + (W \cos \psi_p - U - (V + L) \sin \psi_p) \tan \phi$

Activating force $= W \sin \psi_p + (V + L) \cos \psi_p$

Factor of safety = Resisting force/activating force

Table 5.4b. With external forces.

Resisting force $= cA + (W \cos \psi_p - U - (V + L - B) \sin \psi_p + T \cos \theta) \tan \phi$

Activating force $= W \sin \psi_p + (V + L - B) \cos \psi_p - T \sin \theta$

where: W = sum of vertical loads acting on the wedge specified = weight of wedge + point load Q + surcharge $q * (D - X)$

5.3 SPREADSHEET ANALYSIS OF A SLOPE

PROBLEM: A vertical excavation for a basement is being made adjacent to buildings founded on fill. Some adverse rock joints have been noted and in order for the excavation to proceed, rock anchors are required to retain this wedge.

The discontinuity angle (slope failure plane angle) is 70°, and the height is 7 m for this wedge. Above this wedge fill has been placed during construction of the adjoining building. The fill and building slab exerts a surcharge of 55 kPa and a column with 400 kN is located at the crown of the slope at 1 m from the edge.

In addition the fill is retained by a wall which exerts a lateral load of 32 kN at the crown. This condition is illustrated below.

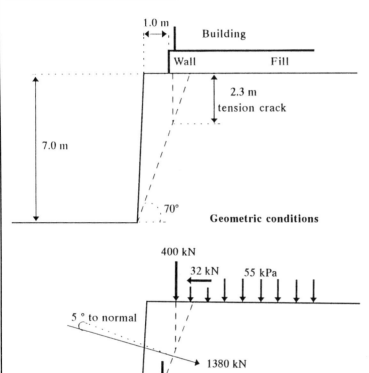

The anchor is to be placed at an angle of 5° to the failure plane normal and with a bolt tension of 1380 kN.

The properties of rock are:
– Cohesion = 15 to 5 kPa,
– Friction angle = 25 to 30 degrees,
– Unit weight of rock = 24 kN/m³.

Due to the fill above the wedge and the pressure of a 2.3 m tension crack at the crown, the slope is expected to experience adverse water conditions.

Evaluate the sensitivity of:

1. The factor of safety to the variation of the restraining (resisting) forces for the range of strength parameters (cohesion and angle of friction) given.

2. The factor of safety for varying water conditions and failure plane angles.

SOLUTION WORKSHEET:
Condition 1: Worksheet 5.1a: Variation of factor of safety with strength parameters.
Condition 2: Worksheet 5.2b: Variation of factor of safety with groundwater conditions and failure plane angles.
The following Lotus functions will be covered:
  – The effect of integrating the worksheet with the graph,
  – The GRAPH AIDS text used in automating the graphical illustration of the results,
  – Use of 'dummy' lines to facilitate graph text entry,
  – The use of multiple embedded (nested) IF functions,
  – The use of trigonometric functions,
  – The conversion from degrees to radian angle for calculations,
  – Use of bar graphs.

Two spreadsheet designs are used to check the sensitivity of the factor of safety with various parameters:
  1. F.S. with varying cohesion and friction angle,
  2. F.S. with varying groundwater conditions and failure plane angles.
Alternatively the worksheets may be set up to evaluate:
  3. F.S. with varying slopes and heights,
  4. F.S. with varying external loading conditions and failure plane angles etc.
The choice of a particular worksheet will be determined by the least known parameter e.g. if the groundwater is considered to be the least known condition and the designer is confident of the loading, geometry, soil strength parameters, then Spreadsheet 2 is chosen.

It should be noted that only three parameters are examined at any one time, i.e. F.S. and two others. This is a physical limitation in simple graphical analysis.

This problem involves a building placed at the top of the slope crown and this imposes a surcharge (due to the floor slab) as well as a point load (column load). The slab has been constructed on filled material which has a retaining wall for support and this acts as a line load at the top of the slope as well as producing a lateral load at the crown of the slope.

Figure 5.4 provides an analysis procedure for this problem.

The three stages for spreadsheet development are:
  1. Set up of data input screen,
  2. Output table,
  3. Graphical output.
The spreadsheet is set up with the headers defining the known parameters. This is done initially at the top of the worksheet. The defined conditions are:
  1. Slope angles and geometry of wedge,
  2. External loading and groundwater conditions, and
  3. Soil properties.

Slope Stability Evaluation

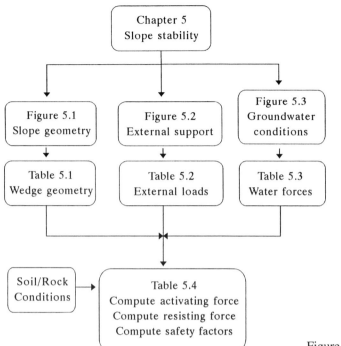

Figure 5.4. Analysis procedure.

WORKSHEET 5.1: SAFETY FACTOR WITH VARYING STRENGTH PARAMETERS

	A	B	C	D	E	F	G	H
1	Stability of a Planar Slope			JOB NO :	**Brisbane**			
2				++++++	++++++	++++++		
3								
4	SLOPE ANG	degrees	Radians		GEOMETRY OF WEDGE			
5	--------				------- --------			
6	Face	90	1.57079		Slope Height – H		7	metres
7	Plane	70	1.22173		Tension crack – z		2.3	metres
8					Water Depth – z(w)		0	metres
9	EXTERNAL LOADING				Tension crack – z(o)		0	metres
10	-------- --------				z(o) = 0 for crack at crown of slope			
11	Surcharge		55	kPa	z(o) > 0 for crack on slope face			
12	Point Load P		400	kN				
13	Dist. from edge Xp		1	metres	Total crown – X		2.5477916	metres
14	Lateral Load L		32	kN	Edge of Crown – D		0.8371315	metres
15	Buttress Load B		0	kN	Plane length – A		5.0016355	metres
16	Bolt Tension T		1380	kN				
17	at Normal Angle of		5	degrees to failure plane				
18	at Normal Angle of		–15	degrees to slope face				
19								
20	GROUNDWATER CONDITIONS			**c**				
21	-------- --------			?????????				
22	Dry Slope : y		Water conditions (a,b or c)					

Worksheet 5.1a. Display screen.

	A	B	C	D	E	F	G	H
22	Dry Slope : y		Water conditions (a,b or c)					
23			a : In Tension Crack Only					
24			b : In Tension crack & sliding surface					
25			c : Saturated slope with heavy recharge					
26								
27								
28	SOIL PROPERTIES		Average	Minimum	Maximum			
29	---------	--------	-------	-------	-------			
30	Cohesion (kPa)		10	5	15			
31	Angle of friction		25	20	30			
32	Unit Wght (kN/cu m)		24	24	24			
33	---------	--------	-------	-------	-------			
34								
35	FORCES (kN / m)					TOTAL LOAD		
36	+++++++++	+++++++++	+++++++	+++++++		+++++++++	+++++++++	
37	Weight of Sliding wedge			190.91		Vertical	731.04	
38	Horizontal water force			25.95		Horizontal	57.95	
39	Uplift Water force			56.43		Uplift	56.43	
40								
41	ACTIVATING Forces (kN) =			586.50	and independent of strength paramaters			
42								
43	Calculation of Slope Safety factor variation with soil properties							
44	+++++++++	+++++++++	+++++++	+++++++	+++++++	+++++++++		

Worksheet 5.1b. Display screen.

	A	B	C	D	E	F	G	H
45		RESISTING		FORCES (kN)				
46	Cohesion		Angle of friction (degrees)					
47	(kPa)	20.0	22.5	25.0	27.5	30.0		
48	+++++++	+++++++	+++++++++++	+++++++++	+++++++++	+++++++++		
49	5.0	576.02	652.09	730.95	813.09	899.06		
50	7.5	588.53	664.59	743.46	825.60	911.56		
51	10.0	601.03	677.09	755.96	838.10	924.07		
52	12.5	613.53	689.60	768.46	850.61	936.57		
53	15.0	626.04	702.10	780.97	863.11	949.07		
54								
55	+++++++	+++++++	+++++++++++	+++++++++	+++++++++	+++++++++		
56			FACTOR OF SAFETY					
57	Cohesion		Angle of friction (degrees)					
58	(kPa)	20.0	22.5	25.0	27.5	30.0		
59	+++++++	+++++++	+++++++++++	+++++++++	+++++++++	+++++++++		
60	5	0.98	1.11	1.25	1.39	1.53		
61	7.5	1.00	1.13	1.27	1.41	1.55		
62	10	1.02	1.15	1.29	1.43	1.58		
63	12.5	1.05	1.18	1.31	1.45	1.60		
64	15	1.07	1.20	1.33	1.47	1.62		
65	+++++++	+++++++	+++++++++++	+++++++++	+++++++++	+++++++++		
66	GRAPH AIDS							
67	Saturated slope with Slope face at			90				
68			ACTIVATING FORCE is independent of the strength paramaters					

Worksheet 5.1c. Display screen.

A job number is used for the worksheet reference and this is the first cell entry.

The highlighted cells illustrated on the Worksheet 5.1 display screen are for the data entry (Chapter 3 for protect and unprotect features).

The spreadsheet is given as Worksheets 5.1a to 5.1c because of the screen shifts required to display the entire worksheet.

Step 1: Data input

The job description is defined in the first row followed by the face and plane angles (in degrees) which are then converted to radians, the spreadsheet calculation format. B7 and B8 are the defined input slope angles in degrees.

Cell (Format)	Display	Cell Entry Formula	Heading/ Formula
C7	1.5707963	@PI*B7/180	Slope face angle (radians)
C8	1.2217305	@PI*B8/180	Slope plane angle (radians)

Worksheet 5.1. Rows 4 to 8: Slope angle input.

E	F	G	H
GEOMETERY OF WEDGE			
----------	----------		
Slope Height – H		7	metres
Tension crack – z		2.3	metres
Water Depth – z(w)		0	metres
Tension crack – z(o)		0	metres
z(o) = 0 for crack at crown of slope			
z(o) > 0 for crack on slope face			
Total crown – X		2.5477916	metres
Edge of Crown – D		0.8371315	metres
Plane length – A		5.0016355	metres

Worksheet 5.1. Rows 4 to 15: Geometry input.

The geometrical definition includes an evaluation of whether the tension crack occurs on the slope face or the crown of the slope. A note has also been included at this stage for ease of use. This acts as a reminder on the $z(0)$ value for the tension crack at the crown of the slope or the slope face.

This cell entry is text and does not calculate or act as any input for the worksheet.

The other geometrical conditions to fully define the wedge geometry are then calculated (Table 5.1 and Figure 5.1). At cells G13, G14, G15 the total crown, edge of crown and plane length respectively are calculated using the trigonome-

Cell (Format)	Display	Cell Entry Formula	Heading/ Formula
G13	2.54779	+G6/@TAN(C7)–G6/@TAN(C6)	Total crown – X : H / tan ψ_f – H / tan ψ_p
G14	0.83715	+G7/@TAN(C7)	Edge of crown – D : z / tan ψ_f
G15	5.0016315	(G6–G7)/@SIN(C7)	Plane Length – A : (H–z) / sin ψ_p

Worksheet 5.1. Calculation of wedge geometry.

tric functions. The Lotus format of the trigonometric functions is @ tan (value in radians).

The referenced cells are:
G6 = slope height
G7 = tension crack
C6 = face angle
C7 = plane angle

Loading and groundwater condition

Data for the external loads is input, with the calculated result of the normal angle to the slope face at cell C18. The cell entry is: + C17 – (B6-B7).

A	B	C	D
EXTERNAL LOADING			
------------	------------		
Surcharge		55	kPa
Point Load P		400	kN
Dist. from edge Xp		1	metres
Lateral Load L		32	kN
Buttress Load B		0	kN
Bolt Tension T		1380	kN
at Normal Angle of		5	degrees to failure plane
at Normal Angle of		–15	degrees to slope face

Worksheet 5.1. Rows 9 to 18: Loading input.

The four groundwater conditions are then defined with a few notes to assist the user for input of the data. These four groundwater conditions are defined in Table 5.3. In this worksheet these conditions are referred to as y, a, b, and c. Based on this input the horizontal (V) and uplift (U) forces are calculated.

A	B	C	D	E
GROUNDWATER CONDITIONS		c		
-----------	----------		????????	
Dry Slope	: y	Water conditions (a,b or c)		
		a : In Tension Crack Only		
		b : In Tension crack & sliding surface		
		c : Saturated slope with heavy recharge		
SOIL PROPERTIES		Average	Minimum	Maximum
-----------	----------	----------	----------	----------
Cohesion (kPa)		10	5	15
Angle of friction		25	20	30
Unit Wght (kN/cu m)		24	24	24
-----------	----------	----------	----------	----------

Worksheet 5.1. Rows 19 to 34: Input of water conditions and soil properties.

The user must choose here one of the four ground water conditions. In this problem condition *c*, which is the worst case, is chosen.

Soil properties
The sensitivity of the model to a range of strength parameters is defined here for evaluation. This may represent the upper and lower bounds expected as well as the average condition. In the tabular computation which follows, the quartile values are also tabulated for evaluation.

Up to this point the known conditions have been input with the 'unknown' soil properties defined in terms of the 'expected' cohesion and friction angle conditions (average values) as well as the minimum and maximum values. This will provide the range for the parametric evaluation.

Step 2: Calculated output

The next step is to produce the output table with the variation of forces and factors of safety in terms of the strength parameters. This involves:

1. Calculation of the activating forces in the first instance, as this is independent of the strength parameters.

2. Calculation of the resisting forces with varying cohesion and frictional components of strength. This variation can subsequently be used to determine the sensitivity of the strength components to the slope stability.

3. The factors of safety are then tabulated as the resisting forces divided by the activating forces.

Activating forces

	A	B	C	D
FORCES (kN / m)				
++++++++++++	++++++++++	++++++++++	++++++++++	
Weight of Sliding wedge				190.91
Horizontal water force				25.95
Uplift Water force				56.43

Worksheet 5.1. Rows 35 to 41: Forces on wedge-output.

The forces are calculated independent of the strength parameters. The formulae given in Tables 5.1 and 5.3 for the weight of wedge and the water forces respectively, are used. The total load is the summation of the forces calculated and the defined parameters in Table 5.3.

The formula entry for the calculated results of the forces and total load is shown.

Cell (Format)	Display	Cell Entry Formula	Heading/ Formula
D37	190.91	0.5*C32* (G6*G13−G14*G7+G9*(G14−G13))	Weight of Sliding wedge/Table 5.1
D38	25.95	@IF(D20="y",0, @IF(D20="a",0.5*9.81*G8*G8, @IF(D20="b",0.5*9.81*G8*G8,0.5*9. 81*G7*G7)))	Horizontal water force/Table 5.3 & Figure 5.3
D39	56.43	@IF(D20="y",0, @IF(D20="a",0, @IF(D20="b",0.5*9.81*G8*G15,0.5*9 .81*G7*G15)))	Uplift Water force/Table 5.3 & Figure 5.3

Worksheet 5.1. Rows 37 to 39: Forces on wedge.

Cell D37: Table 5.1 showed two different formulae for calculation of the weight of the wedge, depending on the location of the tension crack. However, when the tension crack is at the crown of the slope and $z_o = 0$, this equation is simply a subset of the main equation for the tension crack at the slope face. Therefore at cell D37 only the latter equation is used. In algebraic notation this is:

$$\tfrac{1}{2} * \gamma * [HX - DZ + Z_o (D - X)]$$

Cells D38 and D39: The use of text ('y', 'a', 'b', 'c') should be noted in the IF command. This is the criteria which is used to invoke the relevant formula. For the horizontal force calculation (cell D38) there are three conditions which are: dry, partly filled with water and completely filled with water. For the uplift forces, there are two conditions to be applied, $a = b$, and $c = d$.

LOGIC FUNCTION (Horizontal Force)	FORMULA / VALUE
D20 ="y"	0
D20="a"	
D20="b"	0.5*9.81*G8*G8
D20="c"	0.5*9.81*G8*G8 or 0.5*9.81*G7*G7

Worksheet 5.1. Horizontal water forces.

LOGIC FUNCTION (Uplift Force)	FORMULA / VALUE
D20 ="y"	0
D20="a"	
D20="b"	0.5*9.81*G8*G15
D20="c"	0.5*9.81*G7*G15

Worksheet 5.1. Uplift water forces.

The IF. . THEN logic defines the formula to be used.

While the @IF has been used in Chapter 4, the entries shown here are a bit more advanced. The @IF is nested within each other. In this example there are three layers of embedded IF commands to cover the four water conditions. An alternative to the multiple @IF is covered in Chapter 7.

NOTE: For the fully saturated conditions, depth of water (z_w) = depth of tension crack (z). However in the input data z_w (cell G8) was input as zero (a mistake), but the calculation of a fully saturated condition used cell G7 (depth of tension crack) and avoids the possibility of the data input conflict.

Cells G37, G38 and G39: These cells sum the external loads which provide activating forces to the natural loads of the wedge. At Cell G37, the position of the point load is checked to ensure it is within the crown of the slope and not outside the wedge area. The point load is then added to the line load and the surcharge, if it affects the wedge.

The activating forces are calculated from the total load which is the above plus any external loads (surcharge, point loads or lateral loads).

The activating force is independent of the strength parameters and is the component force parallel to the failure plane. This comprises the vertical and horizontal load components. The force component parallel to the failure plane is independent of the uplift force. Table 5.4 gives the formula as:

$$W \sin \psi_p + (V + L - B) \cos \psi_p - T \sin \theta$$

with the required Lotus formula:

```
($G$37*@ sin ($C$7)
+ $G$38*@cos ($C$7)
- $C$16*@ sin ($C$17*@PI/180))
```

The conversion of θ to radians should be noted. The failure plane and face angle were converted to radians earlier in the worksheet.

NOTE: For flexibility of the worksheet the surcharge is taken over the full length of the crown (C11 * G13) – refer cell G37. In reality for this example the surcharge is actually over a distance of $X - X_p$.

The activating force has now been calculated and does not depend on the cohesion or frictional component of strength. This is added as a text note reminder in the worksheet. The resisting force which is dependent on the strength parameters will now be tabled and evaluated.

Cell (Format)	Display	Cell Entry Formula	Heading/ Formula
G37	731.04	@IF(C13<G13, +D37+C12+C11*G13,+D37+C11*G13)	Vertical
G38	57.95	+D38+C14−C15	Horizontal
G39	56.43	+D39	Uplift
D41	586.50	($G$37*@SIN($C$7)+$G$38*@COS($C$7) −$C$16*@SIN($C$17*@PI/180))	ACTIVATING Forces (kN) Table 5.4b

Worksheet 5.1. Rows 37 to 41: Total load.

Resisting forces
The computation table is based on the range of strength parameters defined at the start of the worksheet.

The strength parameters of cohesion and angle of friction are first defined in A49.A53 and B47.F47 respectively.

The resisting forces are then calculated, based on the formula of Table 5.4b.

$$cA + (W \cos \psi_p - U - (V + L - B) \sin \psi_p + T \cos \theta) \tan \phi$$

This is basically a formula entry which uses trigonometric functions and the

A	B	C	D	E	F
Calculation of Slope Safety factor variation with soil properties					
+++++++++++	++++++++++	++++++++++	++++++++++	++++++++++	++++++++++
	R E S I S T I N G		F O R C E S (kN)		
Cohesion		Angle of friction (degrees)			
(kPa)	20.0	22.5	25.0	27.5	30.0
+++++++++++	++++++++++	++++++++++	++++++++++	++++++++++	++++++++++
5.0	576.02	652.09	730.95	813.09	899.06
7.5	588.53	664.59	743.46	825.60	911.56
10.0	601.03	677.09	755.96	838.10	924.07
12.5	613.53	689.60	768.46	850.61	936.57
15.0	626.04	702.10	780.97	863.11	949.07

Worksheet 5.1. Rows 42 to 54: Calculation of resisting forces-output.

mathematical constant PI. Because the trigonometric variables have been converted to radians at the start of the worksheet, the defined slope and plane angles may be input directly. However, the angle of friction value is in degrees and to be used in the caculation the conversion to radians must be made in the trigonometric functions. This is seen as:

`@tan(B47*@PI/180)` ... for the B column

Alternatively the friction angle may have been converted to radians prior to these long formulae in order to make it look less cumbersome and more manageable.

The ABSOLUTE ($) value of all cells referenced in cell B49 was used – except the angle of friction B47. This was then copied across to C, D, E, F and the angles of friction calculation would change to C47, D47, etc. Using the EDIT (F2) while over cell B49, the cohesion A49 was made relative by removing the $ $ and then making the angle of friction $B$47 absolute. This formula was then copied *down* and the cohesion values would then change relative to the row. This is repeated for the other columns.

Cell (Format)	Display	Cell Entry Formula	Heading/ Formula
A49	5	+D30	Cohesion
A50	7.5	(+C30+D30)/2	
A51	10	+C30	
A52	12.5	(+C30+E30)/2	
A53	15	+E30	
B47	20	+$D$31	Angle of friction
C47	22.5	($D$31+$C$31)/2	
D47	25.0	+C31	
etc.		etc.	

B49	576.02	($A$49*$G$15+($G$37*@COS($C$7)
		−$G$39−$G$38*@SIN($C$7)
		+$C$16*@COS($C$17*@PI/180))
		@TAN(B47@PI/180))
C49	652.09	(A49*$G$15+($G$37*@COS($C$7)
		−$G$39−$G$38*@SIN($C$7)
		+$C$16*@COS($C$17*@PI/180))
		@TAN($C$47@PI/180))
etc.		etc.
B50	588.53	(A50*$G$15+($G$37*@COS($C$7)
		−$G$39−$G$38*@SIN($C$7)
		+$C$16*@COS($C$17*@PI/180))
		@TAN($B$47@PI/180))
C50	664.59	(A50*$G$15+($G$37*@COS($C$7)
		−$G$39−$G$38*@SIN($C$7)
		+$C$16*@COS($C$17*@PI/180))
		@TAN($C$47@PI/180))
etc.		etc.
B51	601.03	(A51*$G$15+($G$37*@COS($C$7)
		−$G$39−$G$38*@SIN($C$7)
		+$C$16*@COS($C$17*@PI/180))
		@TAN($B$47@PI/180))
C51	677.09	(A51*$G$15+($G$37*@COS($C$7)
		−$G$39−$G$38*@SIN($C$7)
		+$C$16*@COS($C$17*@PI/180))
		@TAN($C$47@PI/180))
etc.		etc.

RESISTING FORCES of Table 5.4b

Worksheet 5.1. Calculation of resisting forces.

Factor of safety

The last table (Worksheet 5.1c) is the calculation of the factors of safety.

The factors of safety are calculated by the ratio of the variable resisting forces (B49.F53) divided by the constant activating force (D41).

	A	B	C	D	E	F
	++++++++++++	+++++++++++	+++++++++++	+++++++++++	+++++++++++	+++++++++++
			FACTOR OF SAFETY			
	Cohesion		Angle of friction (degrees)			
	(kPa)	20.0	22.5	25.0	27.5	30.0
	++++++++++++	+++++++++++	+++++++++++	+++++++++++	+++++++++++	+++++++++++
	5	0.98	1.11	1.25	1.39	1.53
	7.5	1.00	1.13	1.27	1.41	1.55
	10	1.02	1.15	1.29	1.43	1.58
	12.5	1.05	1.18	1.31	1.45	1.60
	15	1.07	1.20	1.33	1.47	1.62
	++++++++++++	+++++++++++	+++++++++++	+++++++++++	+++++++++++	+++++++++++

Worksheet 5.1. Rows 55 to 65: Calculation of the factor of safety.

No further explanations are provided here, as this is a straight forward formula entry. (After being able to create the table of activating forces as explained above, this table will seem rather simple).

Step 3: Graphical output

For this example a line graph, X-Y graph or bar graph may be used to illustrate the results. In this problem using the XY and line graph would produce the same result. However the XY graph is preferred as this would provide greater flexibility in the event that the average cohesion was not the mean of the maximum and minimum values.

An X-Y graph of resisting forces (Y axis) versus cohesion (X axis) for varying angles of friction can now be developed.

```
/ Graph Type X-Y
                X (A49.A53)              ...Cohesion data range
                A (B49.B53)         ... Range for lowest angle of friction
                B (C49.C53)          ... Range for next angle of friction
                etc.
```

The `Options` facility is then used to define the graph titles as well as the legends of the graph.

```
/ G Options Titles
                First                \A1 {Stability of a planar slope}
                Second                         \E1 {Brisbane}
                X-axis                          Cohesion kPa
                Y-axis                          Forces (kN)
```

Note that the titles may be defined with a cell reference or a text. The use of a cell reference (preceded by \) is useful where the definition of the graph is changing. In this case the site 'Brisbane' – cell E1 is a variable.

The legend must now be specified for the angles of friction. While still in the `Graph Options` menu use:

```
Legend A \B47             ... Friction angle of 20 degrees in this cell
         B \C47           ... Friction angle of 22.5 degrees in this cell
         D \D47                                              ...etc.
         E \E47
         F \F47
```

To this point the basic graph has now been produced. The graph may be seen by pressing the function key `{F10}` or `\Graph View`. The graph provides the following information:
- Plot of forces vs cohesion for varying friction angles,
- Titles, axis labelling and legends.

However to be really useful as a design tool, the comparison between the varying resisting force and the activating force should be shown on the graph.

The activating force is not dependent on the strength parameters (cohesion and friction angle) and therefore would be represented by a straight line (a constant) on the graph. A set of pseudo points must then be established to provide this line. This is first accomplished by putting 'dummy' parameters in a range of cells and then using the F Range setting with the corresponding data range for F.

The dummy line is based on the activating force value $D$41, which is copied to the range of cells G49.G53.

/ Copy from $D$41 to G49.G43

As this line is only for graphical comparison of the variable resisting force with the constant activating force, then the column is hidden from visual display as follows:

/ Range Format Hidden G49.G53

Worksheet 5.1c shows the screen with no display in the G column because of the hidden range format.

The activating force is independent of the strength parameters and this is noted on the graph as a gentle reminder, by inclusion of the data label commands.

Graph Options is used and the range is formatted by having only the lines without the symbols displayed. A straight line is thus produced as shown in Graphsheet 5.1. The following key sequences are used:

/ G F G49.G53 {Data range for dummy line}
   Options Format F Lines {Only line without symbols is displayed for dummy line}
   Options Data-Label F Range A68.C68 {Text range which states 'ACTI-VATING FORCE is independent of the strength parameters'}

Graphsheet 5.1a shows the resulting output of resisting force versus cohesion for varying angle of friction. Knowing the activating forces, one can then plan the remedial measures. The range of strength parameters would guide the design which can compare the effect of operational, minimum, lower quartile strength, etc.

The factor of safety plot is shown next in Graphsheet 5.1b. In order to have both graphs within the same worksheet, it becomes necessary to name the graphs individually. This is accomplished as follows:

/ Graph Name Create {Forces}

The current graph settings are now saved under the graph name of 'Forces'.

A further graph may now be set up as previously but in this instance the factor of safety is plotted against cohesion with varying angles of friction and the sensitivity of the model to changes in strength is evaluated.

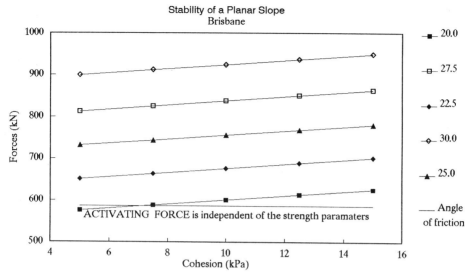

Graphsheet 5.1a. Forces.

As previously, a line graph is used, and is generated as shown:

```
/ Graph Type X-Y
              X (A 60.A64)                    ...Cohesion data range
              A (B60.B64)          ... Range for the lowest angle of friction
              B (C60.C64)           ... Range for the next angle of friction
              etc.
```

The Options facility is then used to define the graph titles and legends.

In order to keep notes on to the graph, the data label key is used to place added information at the top of the screen. This is first accomplished by putting 'dummy' parameters in a range of the cells and the data range for F is then set up.

```
/ G Options Titles
              First                      \A1 {Stability of a planar slope}
              Second                             \E1 {Brisbane}
              X-axis                            Cohesion kPa
              Y-axis                          Factor of Safety
```

The 'dummy' line is created to allow for text notes as:

```
/ G F G60.G64 {Data range for dummy line}
    Options Format F Neither {Neither line nor symbol is displayed for
    dummy line}
    Options Data-Label Range A67.D67
```

To automate the added text facility, the text must be positioned to vary with changing design conditions. The maximum and minimum factors of safety would be in cells F64 and B60 respectively. Therefore this represents the spread of the graph for plotting, and is used for 'creating' the extra space at the top of the graph. The dummy line can then be based (arbitrarily) on {($F$64-$B$60) / 5 + $F$64} (do not read too much into this relationship) in order to create a space just above the top line of the graph. This value is then copied to occupy cells G60.G64.

As these values are only for graph enhancement and have no significance in its value, the column is hidden from visual display as:

```
/ Range Format Hidden G60.G64
```

Worksheet 5.1c shows the screen with no display in the G column so the user is unaware of the 'dummy' creation line.

The text range A67.D67 is a logic function which varies with the data input. In this example the text will show:

  – A saturated slope with heavy discharge."
  – Slope face at (deg)
  – 90

However at these cells A67.C67 the actual entries are:

A67: @IF(D20 = 'y', A21, @IF(D20 = 'a', C23, @IF(D20 = 'b', C24, C25)))
C67: 'Slope face at (degs)
D67: +B7

Design notes are therefore integrated with the graphical output. The slope face

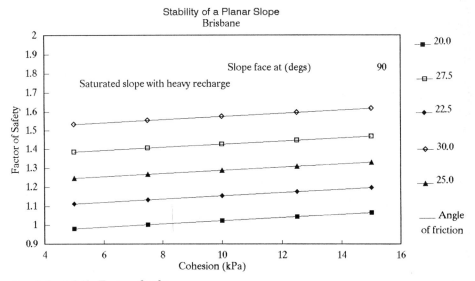

Graphsheet 5.1b. Factor of safety.

angle is also stated. The text line on the graph would vary depending on the conditions defined on data entry. The conditions – Graph and Table – would be automatically updated as the data is input (provided automatic calculation is ON).

> NOTE: Version 3.X allows footnotes to be included in the graph. Version 2.X and 3.X can also be enhanced by using WYSIWYG Add-in.

The graph is now named:

`/ G Name Create {Safety F}`

Graphsheet 5.1b shows the graphical output for this example.

When this worksheet is saved, the last graph used will be displayed when {F10} or / Graph View is used. To view the other named graph use:

`/ Graph Name Use {Forces}`

WORKSHEET 5.2: F.S. WITH VARYING GROUNDWATER CONDITIONS AND FAILURE PLANE ANGLES

	A	B	C	D	E	F	G	H
1	Stability of a Planar Slope			JOB NO :	Brisbane			
2					++++++++ ++++++++++			
3								
4	SLOPE ANGLE				GROUNDWATER CONDITIONS			
5	-------- --------		---------		---------- ------ -			
6	OF	(degrees)	(Radians)		d: Dry Slope Water conditions (a,b,c)			
7	Face	90	1.5707963		a: In Tension Crack Only			
8	Plane avg	70	1.2217304		b: In Tension crack & sliding surface			
9	Min. Plane	60	1.0471975		c: Saturated slope with heavy recharge			
10	Max. Plane	75	1.3089969					
11					GEOMETERY OF WEDGE			
12	SOIL PROPERTIES				---------- ------ - ----------			
13	-------- --------		---------		Slope Height – H 7 metres			
14	Cohesion (kPa)		10		Tension crack – z 2 metres			
15	Angle of friction		25		Water Depth – z(w) 2 metres			
16	Unit Wght(kN/cu m)		24		Tension crack– z(o) 0 below crown			
17								
18	EXTERNAL LOADING				z(o) = 0 for crack at crown of slope			
19	-------- --------		---------		z(o) > 0 for crack on slope face			

Worksheet 5.2a. Display screen.

In some instances the strength parameters (cohesion and angle of friction) may not present the greatest variation, and the designer may be doubtful of the groundwater conditions. This example illustrates the same theory as was covered in Spreadsheet 5.1 but with the groundwater conditions as the variable to be evaluated.

For this example a bar graph is generated and again the spreadsheet is set up with the known variables defined initially. The defined conditions will be the

	A	B	C	D	E	F	G	H
18	EXTERNAL LOADING				z(o) = 0 for crack at crown of slope			
19	-------- -------- ---------				z(o) > 0 for crack on slope face			
20	Surcharge		55	kPa				
21	Point Load P		400	kN				
22	Dist. from edge Xp		1	metres	Total crown − X		2	metres
23	Lateral Load L		32	kN	Edge of Crown − D		0	metres
24	Buttress B		0	kN	Plane length − A		5	metres
25	Bolt Tension T		1380	kN				
26	at Normal Angle of		5	degrees to failure plane				
27	at Normal Angle of		−15	degrees to slope face				
28	Weight of Sliding wedge			190.909	kN/m			
29								
30								
31			GROUNDWATER CONDITIONS					
32	Water Forces		A	B	C	D		
33	-------- -------- --------- ------- ---------- ------ -							
34	Horizontal (H)		19.62	19.62	25.95	0.00		
35	Uplift (U)		0.00	49.07	56.43	0.00		
36								

Worksheet 5.2b. Display screen.

	A	B	C	D	E	F	G
36							
37	TOTAL LOAD (kN/m)		A	B	C	D	
38	-------- -------- -------- -------- -------- ------- -------						
39	Horizontal		51.62	51.62	57.95	32.00	
40	Uplift		0.00	49.07	56.43	0.00	
41	Vertical load (kN/m) =		731.04	and independent of water conditions			
42							
43							
44	Variation of Slope Safety factor with goundwater conditions and slope angle						
45	+++++++++ +++++++++ +++++++++ +++++++++ +++++++++ ++++++++ +++++++						
46	RADIAN	GROUNDWATER CONDITIONS					DEGREE
47	Plane	A	B	C	D		Plane
48	Angle	RESISTING FORCES (kN/m)					Angle
49	+++++++++ +++++++++ +++++++++ +++++++++ +++++++++ ++++++++ +++++++						
50	1.047	840.67	817.79	811.80	848.59		60
51	1.134	813.32	790.44	784.34	821.61		65
52	1.222	785.04	762.16	755.96	793.64		70
53	1.265	770.62	747.74	741.50	779.35		72.5
54	1.309	756.05	733.17	726.89	764.89		75
55	+++++++++ +++++++++ +++++++++ +++++++++ +++++++++ ++++++++ +++++++						

Worksheet 5.2c. Display screen.

slope angles, geometry of wedge, external loading and soil properties. All of the possible groundwater conditions are then used throughout to evaluate the significance of the varying water conditions on the factor of safety. As in Spreadsheet 5.1, the input cells are highlighted. This spreadsheet has basically the same format as Spreadsheet 5.1, but with changes only to the 'variable' input conditions.

	A	B	C	D	E	F	G
56	RADIAN	G R O U N D W A T E R	C O N D I T I O N S				DEGREE
57	Plane	A	B	C	D		Plane
58	Angle	A C T I V A T I N G	F O R C E S (kN/m)				Angle
59	++++++++	++++++++	++++++++	++++++++	++++++++	+++++++	+++++++
60	1.047	538.63	538.63	541.80	528.82		60
61	1.134	564.09	564.09	566.76	555.79		65
62	1.222	584.33	584.33	586.50	577.62		70
63	1.265	592.45	592.45	594.35	586.55		72.5
64	1.309	599.21	599.21	600.85	594.14		75
65							
66							
67	++++++++	++++++++	++++++++	++++++++	++++++++	+++++++	+++++++
68	RADIAN	G R O U N D W A T E R	C O N D I T I O N S				DEGREE
69	Plane	A	B	C	D		Plane
70	Angle	F A C T O R	O F	S A F E T Y			Angle
71	++++++++	++++++++	++++++++	++++++++	++++++++	+++++++	+++++++
72	1.047	1.56	1.52	1.50	1.60		60
73	1.134	1.44	1.40	1.38	1.48		65
74	1.222	1.34	1.30	1.29	1.37		70
75	1.265	1.30	1.26	1.25	1.33		72.5
76	1.309	1.26	1.22	1.21	1.29		75
77	++++++++	++++++++	++++++++	++++++++	++++++++	+++++++	+++++++

Worksheet 5.2d. Display screen.

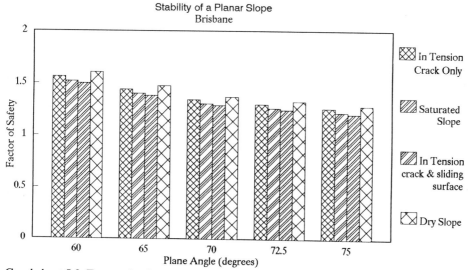

Graphsheet 5.2. Factor of safety.

Worksheet 5.1: Variable cohesion and angle of friction
Worksheet 5.2: Variable groundwater conditions and slope plane

The input of the cell entries follow the same logic as the previous worksheet. The reader may check the formula input from the appendix listings and the worksheet form.

The output Graph 5.2 shows the variation of the four groundwater conditions with factor of safety. A bar graph is produced in this spreadsheet while a line graph is used in Spreadsheet 5.1.

This was done by using:

```
/ Graph Type Bar
```

The bar graph was used in this instance because there are only four definite water conditions for analysis instead of a range of conditions as occurred in the previous example. The graph may be scaled if required by:

```
/ Graph Options Scale Y-Scale Manual Upper
                                      Lower
```

The user has two worksheets with which to carry out an analysis of a planar slope providing a range of conditions for water condition and/or soil strength.

The designer will now have a far greater appreciation of the implication of variations in any of four water conditions.

Similar worksheets may be developed for other parameters.

Stability of a Planar Slope		JOB NO : **Brisbane**
		+++++++++++++++++++++

SLOPE ANGLE   degrees   Radians

GEOMETERY OF WEDGE

	degrees	Radians		
Face	90	1.5708	Slope Height − H	7 metres
Plane	70	1.2217	Tension crack − z	2.3 metres
			Water Depth − z(w)	0 metres
EXTERNAL LOADING			Tension crack − z(o)	0 metres
			z(o) = 0 for crack at crown of slope	
Surcharge		55 kPa	z(o) > 0 for crack on slope face	
Point Load P		400 kN		
Dist. from edge Xp		1 metres	Total crown − X	2.5477794 metres
Lateral Load L		32 kN	Edge of Crown − D	0.8371315 metres
Buttress Load B		0 kN	Plane length − A	5.0016355 metres
Bolt Tension T		1380 kN		
at Normal Angle of		5 degrees to failure plane		
at Normal Angle of		−15.0 degrees to slope face		

GROUNDWATER CONDITIONS   c

?????????

Dry Slope	: y	Water conditions (a,b or c)
		a : In Tension Crack Only
		b : In Tension crack & sliding surface
		c : Saturated slope with heavy recharge

SOIL PROPERTIES

	Average	Minimum	Maximum
Cohesion (kPa)	10	5	15
Angle of friction	25	20	30
Unit Wght (kN/cu m)	24	24	24

```
FORCES (kN / m)                                    TOTAL LOAD
+++++++++++++++++++++++++++++++++++++             ++++++++++++++++
Weight of Sliding wedge              190.91        Vertical        731.04
Horizontal water force                25.95        Horizontal       57.95
Uplift Water force                    56.43        Uplift           56.43

ACTIVATING Forces (kN) =              586.49 and independent of strength parameters

Calculation of Slope Safety factor variation with soil properties
+++++++++++++++++++++++++++++++++++++++++++++++++++++++++++
                    RESISTING  FORCES ( kN )
Cohesion                     Angle of friction (degrees)
        (kPa)    20.0     22.5      25.0      27.5      30.0
+++++++++++++++++++++++++++++++++++++++++++++++++++++++++
         5.0    576.02   652.09    730.95    813.09    899.06
         7.5    588.53   664.59    743.46    825.60    911.56
        10.0    601.03   677.09    755.96    838.10    924.07
        12.5    613.53   689.60    768.46    850.61    936.57
        15.0    626.04   702.10    780.97    863.11    949.07


+++++++++++++++++++++++++++++++++++++++++++++++++++++++++
                    FACTOR OF SAFETY
Cohesion                     Angle of friction (degrees)
        (kPa)    20.0     22.5      25.0      27.5      30.0
+++++++++++++++++++++++++++++++++++++++++++++++++++++++++
          5     0.98     1.11      1.25      1.39      1.53
         7.5    1.00     1.13      1.27      1.41      1.55
         10     1.02     1.15      1.29      1.43      1.58
        12.5    1.05     1.18      1.31      1.45      1.60
         15     1.07     1.20      1.33      1.47      1.62
+++++++++++++++++++++++++++++++++++++++++++++++++++++++++
GRAPH AIDS
Saturated slope with heavy rSlope face          90
                    ACTIVATING  FORCE is independent of the strength parameters
```

Worksheet form 5.1. Data entry and output table. Filename: Slopesoi.

```
Stability of a Planar Slope        JOB NO :Brisbane
                                   ++++++++++++++

SLOPE    ANGLE                     GROUNDWATER CONDITIONS
-------------------------          -----------------------------
   OF    (degrees) (Radians)       d: Dry Slope    Water conditions (a,b,c)
Face          90    1.5708         a: In Tension Crack Only
Plane avg     70    1.2217         b: In Tension crack & sliding surface
Min. Plane    60    1.0472         c: Saturated slope with heavy recharge
Max. Plane    75    1.309
                                   GEOMETERY OF WEDGE
SOIL PROPERTIES                    --------------------------------
-------------------------          Slope Height − H         7 metres
Cohesion (kPa)        10           Tension crack − z        2.3 metres
Angle of friction     25           Water Depth − z(w)       2 metres
Unit Wght(kN/cu m)    24           Tension crack− z(o)      0 below crown
```

```
EXTERNAL LOADING                        z(o) = 0 for crack at crown of slope
------------------------------          z(o) > 0 for crack on slope face
Surcharge                  55 kPa
Point Load  P             400 kN
Dist. from edge Xp          1 metres    Total crown − X        2.5476695 metres
Lateral Load  L            32 kN        Edge of Crown − D      0.8371315 metres
Buttress  B                 0 kN        Plane length − A       5.0016355 metres
Bolt Tension  T          1380 kN
at Normal Angle of          5 degrees to failure plane
at Normal Angle of      −15.0 degrees to slope face
Weight of Sliding wedge        190.9 kN/m
```

GROUNDWATER CONDITIONS

Water Forces	A	B	C	D
Horizontal (H)	19.62	19.62	25.95	0.00
Uplift (U)	0.00	49.07	56.43	0.00

TOTAL LOAD (kN/m)	A	B	C	D
Horizontal	51.62	51.62	57.95	32.00
Uplift	0.00	49.07	56.43	0.00
Vertical load (kN/m) =	731.02 and independent of water conditions			

Variation of Slope Safety factor with goundwater conditions and slope angle
```
++++++++++++++++++++++++++++++++++++++++++++++++++++++++++++++++
```

RADIAN Plane Angle	GROUNDWATER CONDITIONS				DEGREE Plane Angle
	A	B	C	D	
	RESISTING FORCES(kN/m)				
1.047	840.67	817.79	811.80	848.59	60
1.134	813.32	790.44	784.33	821.61	65
1.222	785.04	762.16	755.96	793.64	70
1.265	770.62	747.74	741.49	779.35	72.5
1.309	756.05	733.17	726.89	764.89	75

RADIAN Plane Angle	GROUNDWATER CONDITIONS				DEGREE Plane Angle
	A	B	C	D	
	ACTIVATING FORCES(kN/m)				
1.047	538.62	538.62	541.78	528.81	60
1.134	564.07	564.07	566.74	555.78	65
1.222	584.32	584.32	586.48	577.60	70
1.265	592.43	592.43	594.34	586.53	72.5
1.309	599.20	599.20	600.84	594.12	75

RADIAN	GROUNDWATER CONDITIONS				DEGREE
Plane	A	B	C	D	Plane
Angle	FACTOR OF SAFETY				Angle
1.047	1.56	1.52	1.50	1.60	60
1.134	1.44	1.40	1.38	1.48	65
1.222	1.34	1.30	1.29	1.37	70
1.265	1.30	1.26	1.25	1.33	72.5
1.309	1.26	1.22	1.21	1.29	75

Worksheet form 5.2. Data entry and output table. Filename: Slopegro.

CHAPTER 6

Pile foundations

Pile analysis involves the use of many empirical relationships, with large factors of safety employed because of the variabilities and the consequences of failure. The loading conditions, the type of piles, the method of analysis, the soil conditions and its homogeneity represent some of the variables to be considered.

The theory presented in this chapter covers the analysis for different types of piles, in both cohesive and non-cohesive homogeneous soils, and a two layered soil model with a stiff to hard clay as the underlying strata. Different soil models require a different type of analysis.

Emphasis will be on the use of multiple embedded IF functions integrating what would usually have been separate worksheets for different analysis. The use of the LOOKUP function is shown. This also forms an introduction into the use of the spreadsheet as a knowledge base with the calculations and questions being rule driven, in order to apply the correct method of analysis, and relevant parts of tabulated knowledge being used in the analysis.

6.1 ELEMENTS OF PILE ANALYSIS

Piles are used to transfer loads to an underlying strata. The stress transfer occurs by side friction or end bearing. The ultimate base bearing capacity of the pile (q_b) and the ultimate shaft skin friction or adhesion (q_s) are used to provide the soil design conditions. This would vary with each depth increment.

NOTE: Piles are sometimes referred to as 'friction' or 'end bearing' piles. These represent material idealisations since end bearing piles would have a frictional component, and frictional piles in cohesive soils would have an end bearing component. The division into 'end bearing' and 'friction' piles is simply a convenient terminology to describe the dominant bearing component of the pile.

The actual load carried by the shaft Q_s and base Q_b is given in Table 6.1. The calculation of these loads depends on:
– Soil type,

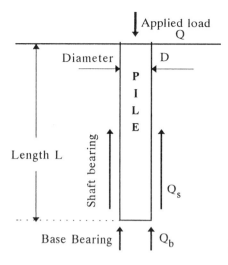

Figure 6.1. Load bearing piles

Table 6.1. Stress transfer.

$$Q_b = q_b * A$$

$$Q_s = q_s * \pi * D * \sum_{l=0}^{l=L} l$$

where: D = diameter; A = area of pile in founding strata; l = length of pile embedment in layer.

– Pile type,
– Embedment depth (Geometric type consideration).
Reference should be made to Tomlinson (1977, 1981) and Meyerhof (1976) for the relevant theory.

6.1.1 *Soil types*

Soils are often analysed as cohesionless or cohesive to determine the bearing capacity of the pile. These are also convenient terms to describe the dominant behaviour characteristic.

The non-cohesive soils are mainly granular material assessed on the basis of the particle size distribution. The cohesive soils are assessed on the stress history of the soil (see Chapter 8 for further explanation). The main soil conditions considered herein are shown in Table 6.2.

Table 6.2. Soil types for pile design.

Non – Cohesive soils (granular)	– Coarse grained sand & gravels – Fine grained sand & non plastic silt	
Cohesive soils (Silts and clays)	– Normally consolidated – Lightly overconsolidated – Highly overconsolidated	Fissured / Non Fissured With / Without Overburden (Homogeneous)

6.1.2 *Pile types*

The piles may be driven or bored. Bored piles are mainly used in cohesive soils or when large end bearing is required. The effect of boring would reduce the soil density or adhesion in the immediate vicinity of the pile and hence a reduction in pile support (friction) from the surrounding soil (for stiff clays only).

> NOTE: Bored piles are also known as cast in situ piles, drilled shafts or drilled piers.

Driven piles are usually used where significant depths are required to achieve the required bearing and/or friction. The main pile types are as follows:
  – Low displacement driven piles,
  – High displacement driven piles,
  – Bored piles (piers).

6.1.3 *Embedment types*

Suitable embedment is required for the strength factors to be applicable. For homogenous soils, the embedment is usually taken as 4 pile diameters in clays and 10 pile diameters in sands. Figure 6.2 shows that as embedment increases the capacity increases to a limiting value.

Where an overburden soil exists over a homogeneous clay, the adhesion coefficient is affected for driven piles by the dragdown effect of the overburden soil for a distance of about 3 pile diameters.

Soft clay over a stiff clay would tend to reduce the adhesion factor of the stiff clay within its upper crust, while for sand over a stiff clay the dragged down sand skin would increase the adhesion value. The effect is shown in Figure 6.3. Design

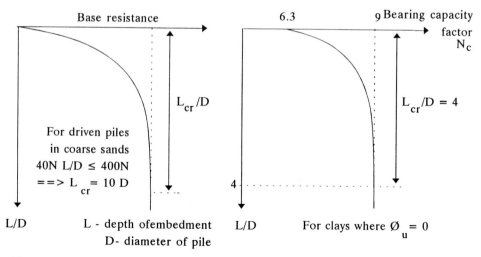

Figure 6.2. Embedment requirement for base resistance in a homogenous soil.

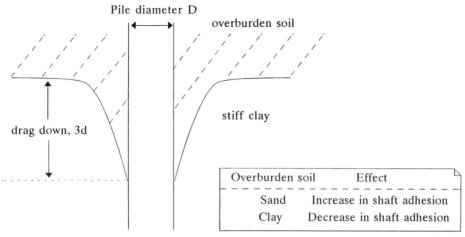

Figure 6.3. Overburden effects on shaft adhesion.

curves are usually used for these conditions to obtain the adhesion coefficient which is a function of the depth of embedment and strength of the soil.

6.1.4 *Factor of safety*

Based on the above conditions/idealisations (soil and pile type) the appropriate numerical relationship is used. A factor of safety is then applied to produce the allowable condition. Typical factors of safety applied are:
1. Shaft factor of safety = 1.5
   Base factor of safety = 3.0
2. Overall factor of safety = 2.5.

The lesser of the two conditions (factored shaft plus base condition, and the overall factor of safety) is then applied in design. The actual factor used is based on the degree of confidence in the relevant elements of design. Lower factors of safety are sometimes applied where pile load tests are carried out.

The design logic is shown in Table 6.3. The main elements to be considered in the spreadsheet analysis of piles would therefore be the variation of soil properties with depth, for a given soil type.

The selected factors of safety are then used to determine the allowable loads and the number of piles required may be estimated from a given pile capacity.

The number of piles required are computed by dividing the total load by the load carrying (structural) capacity of a given pile. This is a first approximation only and the actual number of piles required will depend upon the effects of group efficiency.

The theory for computing the pile capacity in the two main soil types are outlined in the following sections.

Table 6.3. Pile design procedure.

Evaluate design conditions	
Applied condition Piles Bored : Driven	**Existing condition** Soil type Non-Cohesive : Cohesive
Ultimate Capacity of pile with depth Apply factors of safety Compute allowable capacity Compute number of piles required	

6.2 THEORY OF PILES IN NON-COHESIVE SOILS

The ultimate base bearing capacity of piles in non-cohesive (granular) soils can be estimated from the uncorrected N value from the standard penetration test, the depth of embedment, type of soil and type of pile as shown in Table 6.4.

Piles in granular material obtain most of their capacity at the base and are sometimes referred to as end bearing piles.

The empirical relationships of Table 6.4 show the effect of depth of embedment on the ultimate capacity of the pile.

> EXAMPLE: A driven pile in coarse sand will require an embedment of 10 pile diameter before the limiting capacity is reached (400 N). Design approach adopted for this example.

The ultimate shaft bearing capacity of piles driven or bored into non-cohesive soils (gravels, sands and non-plastic silt) can be estimated, based on the pile sizing with the empirical correlations shown in Table 6.5 where (N) = average standard penetration number over the embedded length of the pile within the stratum.

The design process for the piles in non-cohesive soils is shown in Table 6.6.

Table 6.4. End bearing of piles in non-cohesive soils (Meyerhof 1976).

Pile type	Soil type	Ultimate end bearing q_b (kN/m^2)
Driven	Gravels and coarse sands	$40\,N\,L/D \le 400\,N$
Driven	Fine sands and non plastic silt	$40\,N\,L/D \le 300\,N$
Bored	Gravels and coarse sands	$13\,N\,L/D \le 130\,N$
Bored	Fine sands and non plastic silt	$13\,N\,L/D \le 100\,N$

where: N = average Standard Penetration number in the vicinity of pile point (about 10 D above and 4 D below the pile point); L = length of the pile embedded in the granular layer; D = pile diameter.

Table 6.5. Shaft bearing of piles in non-cohesive soils.

Pile type	Pile size	Ultimate shaft bearing q_s (kN/m²)
Driven	High displacement	2(N)
Driven	Low displacement	(N)
Bored	All sizes	(N)

Table 6.6. Pile design in non-cohesive soils.

Select / Action	Output	
Type of pile	Driven	High / low displacement
	Bored	
Non−cohesive soil	Coarse grained Sand and gravel	
	Fine grained sand and non plastic silt	
Input parameters	N (with depth, L)	
	Pile Size D	
Compute	Ultimate base bearing (q_b)	(Table 6.4)
	Ultimate shaft skin friction (q_s)	(Table 6.5)

6.3 THEORY OF PILES IN CLAYS

Piles in clays described in Table 6.2 are considered to be in a saturated homogenous layer, or in a homogeneous clay layer underlying another homogeneous layer (a 2 layer profile).

The ultimate base bearing capacity of piles driven into clays can be determined by the equation of Table 6.7 for undrained conditions.

N_c varies with the sensitivity and deformation characteristics of the clay as shown in Table 6.8.

A value of 9 is however commonly used for driven and bored piles in saturated insensitive and overconsolidated clays in undrained condition. This applies for a minimum of 4 pile diameters embedment as shown in Figure 6.2. An approximate value of 7 is used in this analysis for slightly insensitive and lightly overconsolidated clays.

Table 6.7. End bearing of piles in clays.

q_b (kN/m²) = $c_u N_c$

where: N_c = bearing capacity factor; c_u = undrained cohesive strength (kN/m²).

Table 6.8. Bearing capacity factor N_c with clay type.

Type of clay	Bearing capacity factor N_c
Very sensitive brittle, normally consolidated	5
Slightly sensitive, lightly consolidated	5-10
Insensitive stiff, overconsolidated	10

If a clay is fissured a reduction factor should be used on the laboratory shear strength. A value of 0.75 has been suggested (Skempton 1959) for the base and shaft.

The ultimate shaft capacity of piles driven into clays can be estimated from Table 6.9.

The variation of the adhesion coefficient is given in Table 6.10. In firm to hard clays the values show a significant variation and the selection of the adhesion coefficient value will greatly influence the pile size and number required.

The adhesion coefficient in firm to hard clays depends on the effect of overburden soil, and the depth of embedment. These values are usually read from design graphs, as in Figure 6.4.

As a further limitation on adhesion values, an upper limit of 100 kN/m² should be used for the unit skin friction for bored piers (Skempton 1959).

The design process for piles in cohesive soils is shown in Table 6.11. It must be emphasised again that the actual capacity with depth is dependent on the embedment into that strata as shown in Figure 6.2 for both cohesive and non-cohesive soils.

Table 6.9. Ultimate shaft capacity of piles in clays.

$q_s = \alpha * C_u$ (kN/m²)

where: α = adhesion coefficient which is dependent on the type of clay, the type of pile and the method of installation.

Table 6.10. Adhesion coefficient variation.

Type of pile	Soil type	Adhesion coefficient
Bored	Stiff clays	0.45
	Fissured clays	0.30
Driven	Firm to hard	0.2-1.0
	Soft	1.0

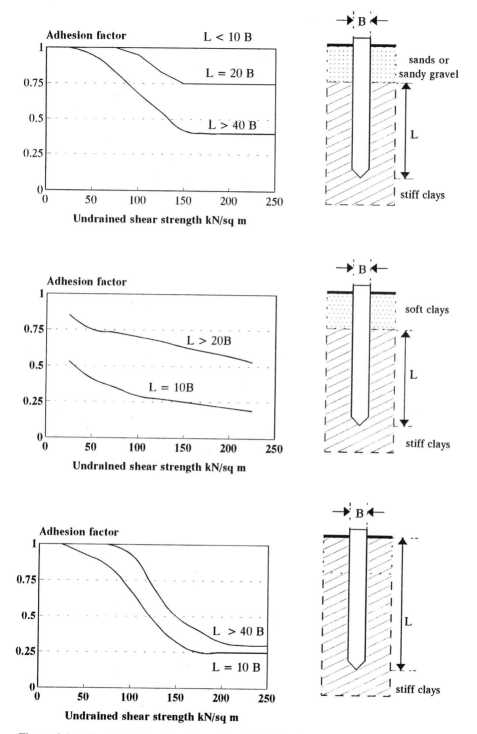

Figure 6.4. Adhesion factors for driven piles (after Tomlinson).

Table 6.11. Pile design for cohesive soils.

Select / Action	Output	
Type of pile	Driven	
	Bored	
Cohesive soil (silts & clays)	Normally consolidated	Fissured / Non – Fissured
	Lightly overconsolidated	Homogeneous / With overlying layer
	Highly overconsolidated	
Input parameters	C_u (with depth, L)	
	Pile size D	
Compute	Ultimate base bearing (q_b)	Bearing capacity factor (Table 6.7 & 6.8)
	Ultimate shaft skin friction (q_s)	Shaft adhesion coefficient (Table 6.9, 6.10 & Fig 6.4)

6.4 SPREADSHEET DESIGN OF PILES IN HOMOGENEOUS SOIL

CONSTRAINT: Combine both of the stated problems into one worksheet so that a user will not require to be moving between spreadsheets.

PROBLEM 1: A 1200 mm diameter bored pile of 1500 kN capacity is proposed at a site where an intact (non-fissured) highly overconsolidated clay was found. The undrained shear strength at 1.5 m intervals was determined.
   Calculate the variation of the ultimate and allowable loads with depth.

PROBLEM 2: A 275 mm square driven pile of 1000 kN capacity is proposed at a site where coarse sand and gravel was found. The Standard Penetration Test was used to evaluate the strength with depth.
   Calculate the variation of the ultimate and allowable loads with depth.

The theory described will be combined into one worksheet to develop the given two problems. Following the development of this worksheet the design graphs of Figure 6.4 will be included into the worksheet.
   The spreadsheet required is therefore an integration, of what would normally have been, two separate worksheets to account for the two distinct analyses of cohesive and non-cohesive soil design conditions.
   The worksheets may be combined in three ways:

Method 1: A front end spreadsheet, which initially defines the type of soil in terms of cohesive/non-cohesive soil. Macros are then used to proceed to the relevant spreadsheet. This may also be accomplished by directly retrieving the known worksheet file for a particular soil type.

Method 2: The condition of cohesive and non-cohesive soils may be examined by using all the above mentioned arguments within one cell with the use of multiple embedded IF functions. The cells contain all possible formulae and the relevant formula is activated by the initial input definitions.

Method 3: One spreadsheet is used as in Method 2, but with the two worksheets found at varying locations. An example of this was provided in Chapter 4 with analysis of different lateral pressures that a wall may experience.

Method 1 is the simplest approach since it essentially involves separate worksheets (Chapter 5 showed an example of using separate worksheets for different analysis). Method 2 has the advantage of the different analysis for the different soil types being under just one worksheet, but with multiple commands in each cell. Method 2 is however more prone to errors in development, because of the complex cell entries.

Method 2 is used here to show the use of the spreadsheet as a rule based system. The use of macros will be illustrated in a subsequent chapter. In both cases the user is unaware of the difference of the logic system since the achieved end result is the same. The different logic which will apply is simply a matter of preference for the spreadsheet designer.

NOTE: An extension of a rule based system would be expert system programs which emulate the problem-solving processes of human experts. This differs from the problem solving (algorithm) type approach of conventional programs where numbers, quantities, and formulae are manipulated in a specific way.

This rule based approach used herein is only able to use IF. . . THEN rules, which for this chapter, differentiates between 6 soil conditions and 3 pile types (18 rules) to apply the knowledge base of formulae to calculate the pile capacity.

Two worksheet screens are shown 6.1 (a..d) and 6.2 (a..d) for cohesive and non-cohesive soils repectively. This is from the same worksheet, but the rule driven nature of the spreadsheet design varies the worksheet form depending on the answers to the initial questions at the top of the form.

EXAMPLE: If (C)ohesive soil was the answer to type of soil, then the next question would relate to the type of consolidation of the clays. Alternatively, if the type of soil had been defined as (N)on-cohesive, then the next question would relate to the grain size of this granular soil.

Subsequently the question path would be based on the answer to the previous question.

The two worksheet screens therefore illustrate the two main paths that the design may take. The case of cohesive soils is illustrated initially in Worksheet 6.1 and for non cohesive soils Worksheet 6.2 is subsequently shown.

In both analysis cases the input parameters are first defined:

Pile conditions:
- Type of pile,
- Pile size (shaft and base),
- Pile shape (round or square),
- Estimated pile load,
- Factors of safety (base, shaft, overall).

Soil conditions (non-cohesive):
- Type of coarse grained soil (coarse sand, gravel, fine sand, non plastic silt),
- Variation of soil strength with depth (N values with depth).

Soil conditions (cohesive):
- Type of fine grained soil (normally, lightly, over consolidated),
- Variation of soil strength with depth (cohesion values with depth).

The worksheet form for pile design is given at the end of the chapter and its cell listing is given in the appendix. The worksheets with explanations of the cell entries are given in this chapter together with the tabular and graphical output.

WORKSHEET 6.1: COHESIVE SOIL

SOLUTION WORKSHEET:
Both worksheets are developed under 1 umbrella but Worksheet 6.1 displays the input/output for cohesive soils while Worksheet 6.2 is for non cohesive soils.
The following Lotus functions will be covered:
- Combining two distinct analyses into one worksheet,
- The use of multiple embedded `IF` commands,
- Follow up questions driven by the answer of preceding questions,
- Using the above techniques, the entire output table (formula and headings) would be driven by the `IF` statements,
- Textural answers instead of numeric values,
- Table lookup,
- Windows,
- Titles.

The display screen for this worksheet is shown as Worksheets 6.1a to 6.1d. The step by step procedure follows below.

DESIGN OF PILES
********** ********

The program considers the following :
          pile type (bored or driven)
          soil type (cohesive or cohesionless)
          variation of pile capacity with length and size

Design Considerations :   Design Conditions
--------- -------   ------ --------

Pile Type: Driven (H)igh or (L)ow displacement, or (B)ored ?   b
                                Pile size (shaft) – d   1200 mm
                                Pile size (Base)   1200 mm
                Pile Shape – (R)ound or (S)quare ?   r
                          Estimated pile load   1500 KN

                      Factor of safety (Bearing)   3
                      Factor of safety (Shaft)   1.5
                      Factor of safety (Overall)   2.5

Worksheet 6.1a. Display screen.

Factor of safety (Bearing)   3
Factor of safety (Shaft)   1.5
Factor of safety (Overall)   2.5

Soil Type – (C)ohesive clays or (N)on–cohesive ?   c
Is the degree of consolidation of the clays   h
(N)ormal, (L)ightly or (H)ighly Overconsolidated, or (F)issured ?
                        Strength paramaters at depth of   1.5 metres
                                    Increment   1.5 metres

Design of Bored piles in highly overconsolidated clays
-------- --------- --------- -------- ---------- ------ ---------

Depth (metres) L	Undrained Shear Strength Cu (kPa)	Embedment Ratio L/d	Adhesion Coefficient of Bored Piles	Max. Bearing Capacity Factor Nc (max)	Bearing Capacity Factor used	Ult Bearing (kN/m/m) Base q(b)

Worksheet 6.1b. Display screen.

Step 1

The first stage is to set up the input form.

    This is the data entry form of the worksheet which drives the rules for the later cells. The question on the pile type (driven or bored), determines the type of equations to be used.

`Pile type – Driven (H)igh or (L)ow displacement, or (B)ored.`

Only the application of driven piles in non-cohesive soils (Table 6.4) is affected by the question of high or low displacement (concrete piles or H piles respectively).

	A	B	C	D	E	F
27	Design of Bored piles in highly overconsolidated clays					
28	--------	---------	---------	--------	----------	-------
29		Undrained	Embedment	Adhesion	Max. Bearing	Bearing
30	Depth	Shear	Ratio	Coefficient	Capacity	Capacity
31	(metres)	Strength		of Bored	Factor	Factor
32	L	Cu (kPa)	L/d	Piles	Nc (max)	used
33	--------	---------	---------	--------	----------	-------
34	1.5	75	1.3	0.45	9.0	6.3
35	3	125	2.5	0.45	9.0	6.3
36	4.5	150	3.8	0.45	9.0	6.3
37	6	100	5.0	0.45	9.0	9.0
38	7.5	125	6.3	0.45	9.0	9.0
39	9	150	7.5	0.45	9.0	9.0
40	10.5	200	8.8	0.45	9.0	9.0
41	12	250	10.0	0.45	9.0	9.0
42	13.5	250	11.3	0.45	9.0	9.0
43	15	300	12.5	0.45	9.0	9.0
44	16.5	350	13.8	0.45	9.0	9.0
45	18	350	15.0	0.45	9.0	9.0
46	--------	---------	---------	--------	----------	-------
47						

Worksheet 6.1c. Display screen.

	A	G	H	I	J	K	L	M
27	Design of Bored piles in							
28	------	--------	------	------	-------	-------	-------	Single
29		Ult Bearing Capacity		Ultimate Load		Total	No of	
30	Depth	(kN/m/m)		(kN)		Allowable	Piles	Pile
31	(metres)	Base	Shaft	Base	Shaft	Load (kN)	required	
32	L	q(b)	q(s)	Q(b)	Q(s)	Q (all)		Capacity
33	------	--------	------	------	-------	-------	-------	--------
34	1.5	471.2	33.8	533.0	190.9	289.5	1	1500
35	3	785.4	56.3	888.3	508.9	558.9	1	1500
36	4.5	942.5	67.5	1065.9	890.6	782.6	1	1500
37	6	900.0	45.0	1017.9	1145.1	865.2	1	1500
38	7.5	1125.0	56.3	1272.3	1463.2	1094.2	1	1500
39	9	1350.0	67.5	1526.8	1844.9	1348.7	1	1500
40	10.5	1800.0	90.0	2035.8	2353.8	1755.8	2	1500
41	12	2250.0	112.5	2544.7	2990.0	2213.9	2	1500
42	13.5	2250.0	112.5	2544.7	3626.2	2468.3	2	1500
43	15	2700.0	135.0	3053.6	4389.6	2977.3	2	1500
44	16.5	3150.0	157.5	3562.6	5280.2	3537.1	3	1500
45	18	3150.0	157.5	3562.6	6170.9	3893.4	3	1500
46	------	--------	------	------	-------	-------	-------	--------
47								

Worksheet 6.1d. Display screen.

For this problem the data entry design conditions are input in column F. The soil strength variation with depth is input at column B.

The logic of Problem 1 is shown in the flowchart, as well as showing some of the other branches which allow for the different analysis path of Problem 2 thereby producing a more flexible worksheet.

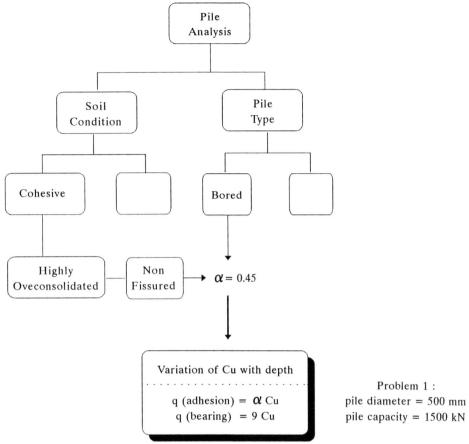

Flowchart 6.1. Design procedure.

Step 2

The next step is also input data on the type of soil. This involves having a series of options to allow for the two soil types. Selection of a particular soil type drives the analysis which will be undertaken.

A	B	C	D	E	F	G
Soil Type – (C)ohesive clays or (N)on–cohesive ?					c	
Is the degree of consolidation of the clays					h	
(N)ormal, (L)ightly or (H)ighly Overconsolidated, or (F)issured ?						
		Strength paramaters at depth of			1.5	metres
				Increment	1.5	metres

Worksheet 6.1. Rows 21 to 26: Input of soil type.

The first prompt defines which of the two types of soil is used in this analysis.

After determining the major soil type, then the next prompt must cater for the further sub-classifications for the specified soil type.

Cell A21 is a text entry which provides the prompt for input in the adjacent cell (F21).

A21: Soil type-(C)ohesive clays or (N)on-cohesive?

Data input at cell F21: C

Cell (Format)	Display	Cell Entry Formula
A22	Is the degree of consolidation of the clays	@IF(F21="c","Is the degree of consolidation of the clays","Is the Non–cohesive soil")
A23	(N)ormal,(L)ightly or (H)ighly Overconsolidated, or(F)issured ?	@IF(F21="n","Gravel and (C)oarse Grained Sand or Non–plastic (S)ilt ?","(N)ormal, (L)ightly or (H)ighly Overconsolidated, or (F)issured ?")

For problem 1 DATA ENTRY at adjacent cell (F22) : **h**

Worksheet 6.1. Soil type defined.

An alternative response is illustrated in Problem 2.

The question is rule driven in order to filter the alternatives for subsequent questions and the formula to be used in the analysis.

The sequence of the logic variables may be input in any order i.e. at cell A23, the cell entry may also be:

@IF(F21="c","(N)ormal, (L)ightly or (H)ighly Overconsolidated, or (F)issured?","Gravel and (C)oarse Grained Sand or Non-plastic (S)ilt?")

Cell (Format)	Display	Cell Entry Formula
A27	Design of Bored Piles in	@IF(F11="b","Design of Bored piles in","Design of Driven piles in")
C27	highly overconsolidated clays	@IF(F21="n", @IF(F22="c","coarse grained sand / gravel","non plastic silt"), @IF(F22="n","normally consolidated clays", @IF(F22="L","lightly overconsolidated clays", @IF(F22="H","highly overconsolidated clays","fissured clays"))))

Worksheet 6.1. Table title defined.

Cells F21 and F22 also predetermine which display prompt screen will be seen. Cell F11 had used 'b' – a bored pile and F21 had used 'c' – a cohesive soil which was further defined in 'F22' as highly overconsolidated. The table header displayed for this example is one of $2 \times 6 = 12$ options available.

The layout of the table also would be based on the data entry.

The header in column B and D has displays 'undrained shear strength' and 'Adhesion coefficient of bored piles' respectively. The header option of cell E32 is similarly defined.

Cell (Format)	Display	Cell Entry Formula
B29	Undrained	@IF(F21="c","Undrained","Standard")
B30	Shear	@IF(F21="c","Shear","Penetration")
B31	Strength	@IF(F21="c","Strength","Test")
B32	Cu(kPa)	@IF(F21="c","Cu (kPa)","(N value)")
C29	Adhesion	@IF(F21="c","Adhesion","Average")
C30	Coefficient	@IF(F21="c","Coefficient","Over")
C31	of Bored	@IF(F21="c",@IF(F11="d","of Driven","of Bored"),"Length")
C32	Piles	@IF(F21="c","Piles","(N aveg)")
E29	Max. Bearing	Max. Bearing
E30	Capacity	Capacity
E31	Factor	Factor
E32	Nc (max)	@IF(F21="c","Nc (max)",@IF(F22="c","40 L/d","13 L/d"))

Worksheet 6.1. Table heading defined.

NOTE: An alternative form will be shown in Chapter 7 using the #AND#, #OR# with the @IF function.

It is seen that each 'question' is based on the preceding answer. If a (D)riven pile had been defined instead, then an alternative table heading and display would be shown.

The remaining header entries of the table are non variable for the calculation of the ultimate bearing capacity and the allowable loads.

Title inferred	Design of driven piles in highly overconsolidated clays	
Table heading inferred	Undrained Shear Strength Cu (kPa)	Adhesion Coefficient of Driven Piles
Data Entry of strength profile		0.3 – 1.0 (refer Table 6.10 – actual value is based on the Cu value input)

Worksheet 6.1. Table heading display.

Title inferred	Design of bored piles in highly overconsolidated clays	
Table heading inferred	Undrained Shear Strength Cu (kPa)	Adhesion Coefficient of Bored Piles
Data Entry of strength profile		0.45 (refer Table 6.10)

Worksheet 6.1. Alternate display.

The adhesion coefficients of Table 6.10 have been 'programmed' into the worksheet (part of the knowledge base), and the values would change depending on the type of soil and pile. In this example, the value of 0.45 is displayed for the case of the bored pile in highly overconsolidated clays.

The use of the series of nested IF functions should be noted. By continuously embedding the IF functions (here and other areas of the worksheet), six soil conditions can be covered in the analysis. This, combined with three pile types, provide eighteen conditions of analysis and hence a more flexible worksheet is produced.

Step 3

The next step is the development of the output table.

Cell (Format)	Display	Cell Entry Formula	Heading/ Formula
A34	1.5	+F24	Depth
A35	3	+A34+$F$25	
A36	4.5	+A35+$F$25	
A37	6	+A36+$F$25	
etc.		etc.	
B34	75	75	Undrained
B35	125	125	Shear
B36	150	150	Strength
B37	100	100	Cu (kPa)
etc.		etc.	

Worksheet 6.1. Rows 34 to 45: Strength variation with depth.

Columns A & B

This is the data entry of the soil parameters with depth. The depth start and increment were defined in cells F24 and F25 and these cells are referenced with the resulting display.

The undrained shear strength with depth is input and is highlighted (unprotected) in the worksheet.

Column C

Column C calculates the embedment ratio (length/diameter) which determines the capacity of the soil at the pile tip. The embedment ratio is defined initially as this will determine the bearing capcity factor to be used.

Cell (Format)	Display	Cell Entry Formula	Heading/ Formula
D34: (F2) PR [W11]	0.45	@IF($F$21="c",@IF($F$11="b", @IF($F$22="f",0.3,0.45),@IF(B34<25,1,@I F(B34>100,0.3,0.5))),@AVG($B$34..B34))	Adhesion Coeff. of Bored Piles
D35: (F2) PR [W11]	0.45	@IF($F$21="c",@IF($F$11="b", @IF($F$22="f",0.3,0.45),@IF(B35<25,1,@I F(B35>100,0.3,0.5))),@AVG($B$34..B35))	
D36: (F2) PR [W11]	0.45	@IF($F$21="c",@IF($F$11="b", @IF($F$22="f",0.3,0.45),@IF(B36<25,1,@I F(B36>100,0.3,0.5))),@AVG($B$34..B36))	
etc.		etc.	
E34: (F1) PR [W13]	9.0	@IF($F$21="c",@IF($F$22="n",5, @IF($F$22="l",7,@IF($F$22="h",9,0.75*9))),@IF($F$11="b",13*C34,40*C34))	Maximum Bearing Capacity factor Nc (Max)
E35: (F1) PR [W13]	9.0	@IF($F$21="c",@IF($F$22="n",5, @IF($F$22="l",7,@IF($F$22="h",9,0.75*9))),@IF($F$11="b",13*C35,40*C35))	
E36: (F1) PR [W13]	9.0	@IF($F$21="c",@IF($F$22="n",5, @IF($F$22="l",7,@IF($F$22="h",9,0.75*9))),@IF($F$11="b",13*C36,40*C36))	
etc.		etc.	
F34: (F1) PR [W9]	6.3	@IF($F$21="c", @IF(C34>4,E34,2*@PI),@IF($F$11="b",@ IF($F$22="c",@MIN(E34,130),@MIN(E34, 100)),@IF($F$22="c",@MIN(E34,400),@M IN(E34,300))))	Bearing capacity factor Used
F35: (F1) PR [W9]	6.3	@IF($F$21="c", @IF(C35>4,E35,2*@PI),@IF($F$11="b",@ IF($F$22="c",@MIN(E35,130),@MIN(E35, 100)),@IF($F$22="c",@MIN(E35,400),@M IN(E35,300))))	
F36: (F1) PR [W9]	6.3	@IF($F$21="c", @IF(C36>4,E36,2*@PI),@IF($F$11="b",@ IF($F$22="c",@MIN(E36,130),@MIN(E36, 100)),@IF($F$22="c",@MIN(E36,400),@M IN(E36,300))))	
etc.		etc.	

Worksheet 6.1. Rows 34 to 45: Adhesion and bearing capacity factors.

Column D

Column D provides the adhesion coefficient based on the type of cohesive soil previouisly defined. If the soil had been granular then an average N value would have been displayed.

Column E

Column E calculates the maximum bearing capacity factor applicable based on the embedment ratio.

Column F

Column F calculates the bearing capacity factor used based on the embedment ratio and limiting values.

Cell (Format)	Display	Cell Entry Formula	Heading/ Formula
G34: (F1) PR [W10]	471.2	+F34*B34	Ult.
G35: (F1) PR [W10]	785.4	+F35*B35	Bearing
G36: (F1) PR [W10]	942.5	+F36*B36	Capacity
G37: (F1) PR [W10]	900.0	+F37*B37	(Base)
etc.		etc.	
H34: (F1) PR [W10]	33.8	@IF($F$21="c",B34*D34,	Ult.
		@IF($F$11="h",2*D34,D34))	Bearing
H35: (F1) PR [W10]	56.3	@IF($F$21="c",B35*D35,	Capacity
		@IF($F$11="h",2*D35,D35))	(Shaft)
H36: (F1) PR [W10]	67.5	@IF($F$21="c",B36*D36,	
		@IF($F$11="h",2*D36,D36))	
H37: (F1) PR [W10]	45.0	@IF($F$21="c",B37*D37,	
		@IF($F$11="h",2*D37,D37))	
etc.		etc.	

Worksheet 6.1. Rows 34 to 45: Ultimate bearing capacity.

Columns G & H

Columns G & H calculate the ultimate base and shaft bearing capacity respectively.

Columns I & J

Columns I & J then calculate the base load at the given depth with respect to the area.

   All the above involve the use of multiple IF logic functions to cover all the possibilities as given in the pile theory at the start of the chapter.

Column K

Column K requires that the minimum factor of safety be applied, since the two conditions (base and shaft factored, and overall factor of safety) must be considered. The function @Min(...) is used to apply the lesser of the two calculated conditions.

Columns L

Column L uses the @INT(...) function for determining the number of piles. A pile must be a whole number and by using the integer value then the fractional part is truncated. The integer number is increased by 1 in order to consider any 'fraction' of a pile as a whole number.

Graphical output
The model can be examined as a variation of ultimate bearing capacity (base, shaft, total) or allowable bearing capacity with depth in order to assess a suitable bearing depth. Alternatively the variation of bearing capacities with factors of safety can be examined at a specific depth in order to assess an acceptable factor of safety.

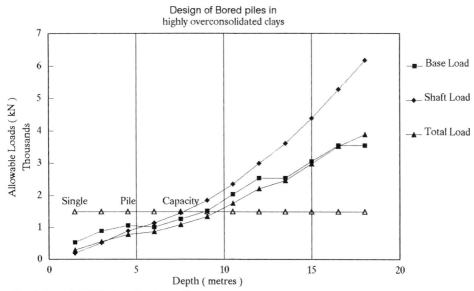

Graphsheet 6.1. Pile in cohesive soil.

WORKSHEET 6.2: NON-COHESIVE SOIL

The worksheets for a non-cohesive pile analysis is now shown to view an alternative selection path within the same worksheet.

Questions to define the type of analysis	What type of pile is proposed –(D)riven or (B)ored ?	D
	Soil type – (C)ohesive clays or (N)on–cohesive ?	N
	Gravel and (C)oarse grained sand or (N)on plastic silt ?	C
Title heading of output table	Design of Driven piles in coarse grained sand/ gravel	

Worksheet 6.2. Design options selected.

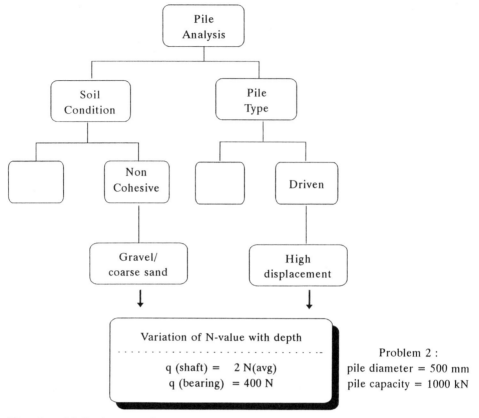

Flowchart 6.2. Design procedure.

	A	B	C	D	E	F	G
1				**DESIGN OF PILES**			
2				********* *********			
3							
4	The program considers the following :						
5			pile type (bored or driven)				
6			soil type (cohesive or cohesionless)				
7			variation of pile capacity with length and size				
8							
9	Design Considerations :					Design Conditions	
10	-------- --------					------- -------	
11	Pile Type: Driven (H)igh or (L)ow displacement, or (B)ored ?					h	
12				Pile size (shaft) – d		275	mm
13				Pile size (Base)		275	mm
14			Pile Shape – (R)ound or (S)quare ?			s	
15				Estimated pile load		1000	KN
16							
17			Factor of safety (Bearing)			3	
18			Factor of safety (Shaft)			1.5	
19			Factor of safety (Overall)			2.5	

Worksheet 6.2a. Display screen.

Worksheets 6.1 and 6.2 should be compared for the logic structure. The logic would vary with the questions as shown.

The display for the worksheet will therefore be quite different, while the format and formulae of the worksheet will be the same.

The user now need refer only to 1 spreadsheet (FILE name: `Pile`) and be presented with 1 worksheet form which will cover all of the alternatives covered in this chapter. The alternatives covered in this chapter have been 'programmed' into the cells with multiple `IF` functions.

	A	B	C	D	E	F	G
16							
17			Factor of safety (Bearing)			3	
18			Factor of safety (Shaft)			1.5	
19			Factor of safety (Overall)			2.5	
20							
21	Soil Type – (C)ohesive clays or (N)on–cohesive ?					n	
22	Is the Non–cohesive soil					c	
23	Gravel and (C)oarse Grained Sand or Non–plastic (S)ilt ?						
24				Strength paramaters at depth of		1.5	metres
25					Increment	1.5	metres
26							
27	Design of Driven piles in coarse grained sand / gravel						
28	---------	---------	---------	-------	----------	-------	--------
29		Standard	Embedment	Average	Max. Bearing	Bearing	Ult Bearing
30	Depth	Penetration	Ratio	Over	Capacity	Capacity	(kN/m/m)
31	(metres)	Test		Length	Factor	Factor	Base
32	L	(N value)	L/d	(N aveg)	40 L /d	used	q(b)

Worksheet 6.2b. Display screen.

	A	B	C	D	E	F
27	Design of Driven piles in coarse grained sand / gravel					
28	---------	---------	---------	-------	----------	-------
29		Standard	Embedment	Average	Max. Bearing	Bearing
30	Depth	Penetration	Ratio	Over	Capacity	Capacity
31	(metres)	Test		Length	Factor	Factor
32	L	(N value)	L/d	(N aveg)	40 L /d	used
33	---------	---------	---------	-------	----------	-------
34	1.5	6	5.5	6.00	218.2	218.2
35	3	15	10.9	10.50	436.4	400.0
36	4.5	18	16.4	13.00	654.5	400.0
37	6	12	21.8	12.75	872.7	400.0
38	7.5	23	27.3	14.80	1090.9	400.0
39	9	27	32.7	16.83	1309.1	400.0
40	10.5	32	38.2	19.00	1527.3	400.0
41	12	38	43.6	21.38	1745.5	400.0
42	13.5	43	49.1	23.78	1963.6	400.0
43	15	55	54.5	26.90	2181.8	400.0
44	16.5	56	60.0	29.55	2400.0	400.0
45	18	48	65.5	31.08	2618.2	400.0
46	---------	---------	---------	-------	----------	-------
47						

Worksheet 6.2c. Display screen.

	Design of Driven piles in						
							Single
		Ult Bearing Capacity		Ultimate Load		Total	No of	
Depth		(kN/m/m)		(kN)		Allowable	Piles	Pile
(metres)		Base	Shaft	Base	Shaft	Load (kN)	required	
L		q(b)	q(s)	Q(b)	Q(s)	Q (all)		Capacity
1.5		1309.1	12.0	99.0	19.8	46.2	1	1000
3		6000.0	21.0	453.8	54.5	187.6	1	1000
4.5		7200.0	26.0	544.5	97.4	246.4	1	1000
6		4800.0	25.5	363.0	139.4	201.0	1	1000
7.5		9200.0	29.6	695.8	188.3	353.6	1	1000
9		10800.0	33.7	816.8	243.8	424.2	1	1000
10.5		12800.0	38.0	968.0	306.5	509.8	1	1000
12		15200.0	42.8	1149.5	377.1	610.6	1	1000
13.5		17200.0	47.6	1300.8	455.5	702.5	1	1000
15		22000.0	53.8	1663.8	544.3	883.2	1	1000
16.5		22400.0	59.1	1694.0	641.8	934.3	1	1000
18		19200.0	62.2	1452.0	744.4	878.5	1	1000

Worksheet 6.2d. Display screen.

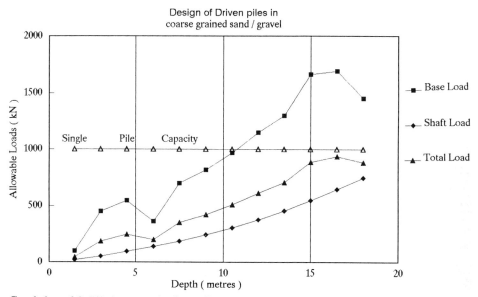

Graphsheet 6.2. Pile in non cohesive soil.

In a topic as diverse as this, the user will find many other aspects that may be programmed into the worksheet. The more alternatives that are used, the wider the knowledge base of the worksheet and therefore the more expert the spreadsheet becomes.

WORKSHEET 6.3: TWO LAYERED SOIL PROFILE

As described in earlier sections the spreadsheet design was for a single pile in a homogenous soil. The option for two layers in an overconsolidated clay at the lower strata can affect the adhesion coefficients considerably. The information shown in Figure 6.4 is developed into a tabled knowledge base using the Lotus @VLookup function to automatically scan a table for the required value.

Lotus has two Lookup functions for looking through the contents of a range and picking a match for a given criteria. The @HLookup and @VLookup return the contents of a cell in a table based on a row and column match respectively.

These functions are used as:

@HLookup (key, range, row-offset).
@VLookup (key, range, column-offset)

A series of nested @IF funtions could have been used to accomplish the same task, but the Lookup function is a more compact equation for automatic scanning of a table.

The worksheet is for a two layered soil profile where driven piles are used. However while the analysis is confined to a clay profile, the flexibility remains to analyse a single layer soil profile or a bored pile instead of a driven pile.

This analysis therefore covers four model conditions:

1. Granular over clay (driven pile),
2. Soft clay over clay (driven pile),
3. Clay only (driven pile),
4. Clay only (bored pile).

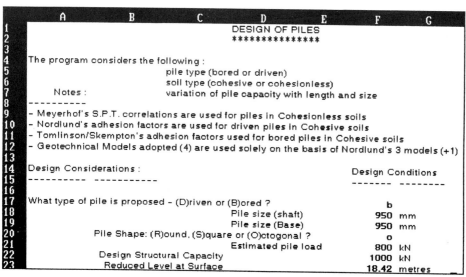

Worksheet 6.3a. Display screen.

The five worksheet display screens are shown. Displays a, b and c input data while the other screens display the calculated results based on the input parameters and model adopted.

Display screen f is the input of the chart data of Figure 6.4 for the three models, strengths and embedment ratio. Based on this tabled data then another table is created to assess the applicable adhesion factor as seen in Worksheet 6.3g. This adhesion is then used in the calculations of Worksheet 6.3c and 6.3d.

A check is first made on the layer to determine whether the depth being analysed falls within or below the sand/gravel thickness.

	A	B	C	D	E	F
24						
25			Factor of safety (Bearing)			3
26			Factor of safety (Shaft)			1.5
27			Factor of safety (Overall)			2.5
28						
29	Soil Type – (C)ohesive clays or (N)on-cohesive ?					c
30	Is the degree of consolidation of the clays					f
31	(N)ormal, (L)ightly or (H)ighly Overconsolidated, or (F)issured ?					
32	----------					
33	Geotechnical Model is considered one layered – Use Model 4					
34	----------	-----------	----------	---------	----------	------- ---
35	Geotechnical Model		|		Model 1 :	Sand/gravel over Stiff Clay
36	1		|		Model 2 :	Soft Clay over Stiff Clay
37	Sand/Gravel thickness		|		Model 3 :	Only Stiff Clay
38	1.8	metres	|		Model 4 :	One Layered Soil Strata
39	----------	-----------	----------	---------	----------	------- ---
40						

Worksheet 6.3b. Display screen.

	A	B	C	D	E	F	G	H
41	Design of Bored piles in fissured clays							
42	-------	----------	--------	--------	--------	------	------	------
43			Undrained	Adhesion	Ult Bearing capacity		Ultimate Load	
44	Depth	Reduced	Shear	Coefficient	(kN/m/m)		(kN)	
45	(metres)	Level	Strength	of Bored	Base	Shaft	Base	Shaft
46		(metres)	Cu (kPa)	Piles	q(b)	q(s)	Q(b)	Q(s)
47	-------	----------	--------	--------	--------	------	------	------
48	0	18.42	0.00					
49	1.8	16.62	0	0.30	0.0	0.0	0.0	0.0
50	4.42	14	100	0.30	600.0	30.0	425.3	234.6
51	8	10.42	200	0.30	1200.0	60.0	850.6	875.7
52	8	10.42	450	0.30	2700.0	135.0	1913.8	875.7
53	9	9.42	450	0.30	2700.0	135.0	1913.8	1278.6
54	9.5	8.92	450	0.30	2700.0	135.0	1913.8	1480.0
55	10	8.42	450	0.30	2700.0	135.0	1913.8	1681.5
56	10.5	7.92	450	0.30	2700.0	135.0	1913.8	1882.9
57	11	7.42	450	0.30	2700.0	135.0	1913.8	2084.4
58	11.5	6.92	450	0.30	2700.0	135.0	1913.8	2285.8
59	-------	----------	--------	--------	--------	------	------	------
60	Borehole 2							
61								

Worksheet 6.3c. Display screen.

If the depth is below the sand/gravel layer, the table illustrated in display screen 6.3f is then scanned, using the @VLookup, to determine the adhesion factor corresponding to the shear strength value – for all embedment ratios.

These values are then automatically tabled, as seen in display screen 6.3g in the range D89 to H99. At column I the actual value to be used in the calculation is determined using an @IF function, with the @HLookup nested within.

The geotechnical model graph adapted can be set up to display Graphsheet 6.3a. The other grapsheets follow from the previous worksheets developed.

	A	I	J	K	L	M	N	O	P	Q	R
41	Design of Bored piles in										
42	-------	-------	-------	------	-----	------	------				
43		Total							Check on embedment		
44	Depth	Allowable	No of	Pile	Percentage Carried by				L/D		Ult Brg Cap
45	(metres)	Load (kN)	Piles		Base	Shaft	Total	for that			Base
46		Q (all)	required	Capacity			Pile	layer	Nc		q(b)
47	-------	-------	-------	------	-----	------	------	----	---	------	
48	0										
49	1.8	0.0	1	800	ERR	ERR	0.0%	1.9	7.6	0	
50	4.42	264.0	1	800	64.5%	35.5%	33.0%	2.8	8.3	830	
51	8	690.5	1	800	49.3%	50.7%	86.3%	3.8	8.8	1760	
52	8	1115.8	2	800	68.6%	31.4%	139.5%	0.0	6.3	2827.4	
53	9	1277.0	2	800	59.9%	40.1%	159.6%	1.1	9.0	4050	
54	9.5	1357.5	2	800	56.4%	43.6%	169.7%	0.5	9.0	4050	
55	10	1438.1	2	800	53.2%	46.8%	179.8%	0.5	9.0	4050	
56	10.5	1518.7	2	800	50.4%	49.6%	189.8%	0.5	9.0	4050	
57	11	1599.3	2	800	47.9%	52.1%	199.9%	0.5	9.0	4050	
58	11.5	1679.9	3	800	45.6%	54.4%	210.0%	0.5	9.0	4050	
59	-------	-------	-------	------	-----	------	------	----	---	------	
60	Borehole 2										

Worksheet 6.3d. Display screen.

	A	B	C	D	E	F	G	H
61								
62	Geotechnical Model is considered one layered							
63	+++++++	+++++++++	++++++++	+++++++	++++++++++			
64								
65	Adhesion Factors for Driven Piles (after Nordlund)							
66	-------	---------	--------	---------	-----	------		
67	Geotechnical Model				Model 1 Sand/gravel over Stiff Clay			
68	1				Model 2 Soft Clay over Stiff Clay			
69	Sand/Gravel thickness				Model 3 Only Stiff Clay			
70	1.8	metres			Model 4 Not applicable here			
71	-------	---------	--------	---------	-----	------		
72	Undrained Shear			Adhesion factor at depth embedded into stiff clay				
73		Strength		-------	---------	--------------		
74		Cu (kPa)		L <= 10 B	15 B	20 B	30 B . > 40 B	
75	---------	--------	-------	---------	-----	------	------	
76								
77		50		1	1	1	0.95	0.9
78		100		1	0.95	0.9	0.75	0.7
79		150		1	0.9	0.75	0.55	0.35
80		200		1	0.9	0.75	0.55	0.35
81								

Worksheet 6.3e. Display screen.

	L	M	N	O	P	Q	R
60							
61					Brg Capacity factor		
62					L/D	Nc	
63					--------	--------	
64					0	6.3	
65					1	7.6	
66					2	8.3	
67					3	8.8	
68					4	9.0	
69							

Worksheet 6.3f. Display screen.

	A	B	C	D	E	F	G	H	I	J
81										
82	------	------	---------------	----	-----	-----	-----	-----		
83	Table is for a Geotechnical Model of Sand/gravel over Stiff Clay									
84	------	------	---------------	----	-----	-----	-----	-----	---	-------
85		Depth	Undrained Shear			Adhesion factor at depth embedded into stiff clay				
86	Depth	(metres)	Strength -----------	-----	-----	---------------				
87	Increment		Cu (kPa)	9.50	14.25	19.00	28.50	38.00	Design Value	
88	------	------	---------------	----	-----	-----	-----	-----	---	-------
89	0	0	0.00001	0	0	0	0	0	0	
90	1	1.8	0	0	0	0	0	0	0	
91	2	4.42	100	1	0.95	0.9	0.75	0.7	1	
92	3	8	200	1	0.9	0.75	0.55	0.35	1	
93	4	8	450	1	0.9	0.75	0.55	0.35	1	
94	5	9	450	1	0.9	0.75	0.55	0.35	1	
95	6	9.5	450	1	0.9	0.75	0.55	0.35	1	
96	7	10	450	1	0.9	0.75	0.55	0.35	1	
97	8	10.5	450	1	0.9	0.75	0.55	0.35	1	
98	9	11	450	1	0.9	0.75	0.55	0.35	1	
99	10	11.5	450	1	0.9	0.75	0.55	0.35	1	
100	------	------	---------------	----	-----	-----	-----	-----	---	
101										
102										

Worksheet 6.3g. Display screen.

The reader should go through the listing for the development of this worksheet.

Analysis and design of piled foundations is a very broad topic with many specialist books written on the subject. Some of the other factors to be considered include:

– The effect of negative friction. This occurs when a pile is embedded in a stratum overlain by more compressible clay or fill.

– The effect of pile groups. Increased capacity for driven piles in granular materials.

– The effect of belling. The case of the enlarged bases on bored piles in a fissured clay would have its effective shaft length reduced by about two pile diameters. This occurs because of the tendency of the soil directly above the under reamed base to slip down. It is prudent to avoid inclusion of this length in the assessment of the pile adhesion as shown in Figure 6.5. Therefore while there is an increase in bearing, the shaft friction has decreased.

Further refinement of the worksheets covered can incorporate the above factors to provide a more flexible worksheet.

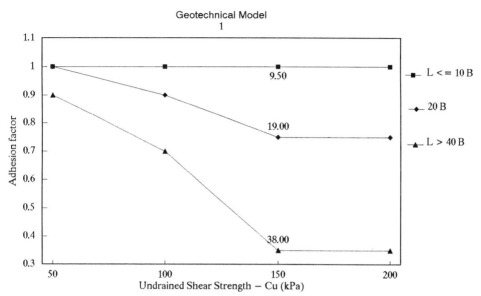

Graphsheet 6.3a. Geotechnical model adopted.

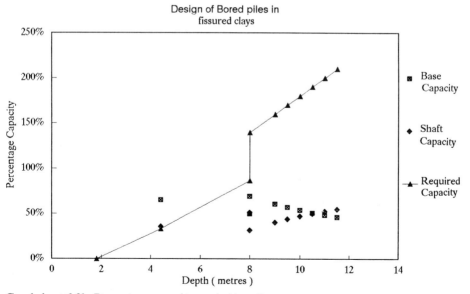

Graphsheet 6.3b. Percentage capacity carried by pile.

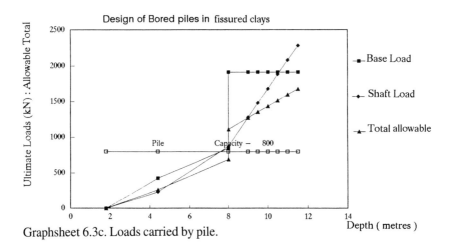

Graphsheet 6.3c. Loads carried by pile.

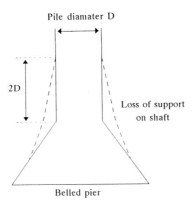

Figure 6.5. Adhesion loss at belled pier.

Design of Bored piles in highly overconsolidated clays

Depth (metres) L	Undrained Shear Strength Cu (kPa)	Embedment Ratio L/d	Adhesion Coefficient of Bored Piles	Max. Bearing Capacity Factor Nc (max)	Bearing Capacity Factor used
1.5	75	1.3	0.45	9.0	6.3
3	125	2.5	0.45	9.0	6.3
4.5	150	3.8	0.45	9.0	6.3
6	100	5.0	0.45	9.0	9.0
7.5	125	6.3	0.45	9.0	9.0
9	150	7.5	0.45	9.0	9.0
10.5	200	8.8	0.45	9.0	9.0
12	250	10.0	0.45	9.0	9.0
13.5	250	11.3	0.45	9.0	9.0
15	300	12.5	0.45	9.0	9.0
16.5	350	13.8	0.45	9.0	9.0
18	350	15.0	0.45	9.0	9.0

DESIGN OF PILES
****

The program considers the following :
- pile type (bored or driven)
- soil type (cohesive or cohesionless)
- variation of pile capacity with length and size

Design Considerations :   Design Conditions

Pile Type: Driven (H)igh or (L)ow displacement, or (B)ored ?   b
                                 Pile size (shaft) – d   1200 mm
                                 Pile size (Base)   1200 mm
               Pile Shape – (R)ound or (S)quare ?   r
                             Estimated pile load   1500 KN

                Factor of safety (Bearing)   3
                Factor of safety (Shaft)   1.5
                Factor of safety (Overall)   2.5

Soil Type – (C)ohesive clays or (N)on–cohesive ?   c
Is the degree of consolidation of the clays   h
(N)ormal, (L)ightly or (H)ighly Overconsolidated, or (F)issured ?
                       Strength paramaters at depth of   1.5 metres
                                   Increment   1.5 metres

Ult Bearing Capacity (kN/m/m)		Ultimate Load (kN)		Total Allowable Load (kN)	No of Piles required	Single Pile
Base q(b)	Shaft q(s)	Base Q(b)	Shaft Q(s)	Q (all)		Capacity
471.2	33.8	533.0	190.9	289.5	1	1500
785.4	56.3	888.3	508.9	558.9	1	1500
942.5	67.5	1065.9	890.6	782.6	1	1500
900.0	45.0	1017.9	1145.1	865.2	1	1500
1125.0	56.3	1272.3	1463.2	1094.2	1	1500
1350.0	67.5	1526.8	1844.9	1348.7	1	1500
1800.0	90.0	2035.8	2353.8	1755.8	2	1500
2250.0	112.5	2544.7	2990.0	2213.9	2	1500
2250.0	112.5	2544.7	3626.2	2468.3	2	1500
2700.0	135.0	3053.6	4389.6	2977.3	2	1500
3150.0	157.5	3562.6	5280.2	3537.1	3	1500
3150.0	157.5	3562.6	6170.9	3893.4	3	1500

Worksheet form 6.1. Data entry and output table. Filename: Pilecoh.

DESIGN OF PILES
****

The program considers the following :
> pile type (bored or driven)
> soil type (cohesive or cohesionless)
> variation of pile capacity with length and size

Design Considerations :	Design Conditions

Pile Type: Driven (H)igh or (L)ow displacement, or (B)ored ?   **h**

Pile size (shaft) — d	275 mm
Pile size (Base)	275 mm
Pile Shape — (R)ound or (S)quare ?	s
Estimated pile load	1000 KN
Factor of safety (Bearing)	3
Factor of safety (Shaft)	1.5
Factor of safety (Overall)	2.5

Soil Type — (C)ohesive clays or (N)on−cohesive ?   **n**
Is the Non−cohesive soil   **c**
Gravel and (C)oarse Grained Sand or Non−plastic (S)ilt ?

Strength paramaters at depth of	1.5 metres
Increment	1.5 metres

Design of Driven piles in coarse grained sand / gravel

Depth (metres) L	Standard Penetration Test (N value)	Embedment Ratio L/d	Average Over Length (N aveg)	Max. Bearing Capacity Factor 40 L /d	Bearing Capacity Factor used
1.5	6	5.5	6.00	218.2	218.2
3	15	10.9	10.50	436.4	400.0
4.5	18	16.4	13.00	654.5	400.0
6	12	21.8	12.75	872.7	400.0
7.5	23	27.3	14.80	1090.9	400.0
9	27	32.7	16.83	1309.1	400.0
10.5	32	38.2	19.00	1527.3	400.0
12	38	43.6	21.38	1745.5	400.0
13.5	43	49.1	23.78	1963.6	400.0
15	55	54.5	26.90	2181.8	400.0
16.5	56	60.0	29.55	2400.0	400.0
18	48	65.5	31.08	2618.2	400.0

NOTE: For large worksheets such as this pile example where screen shifts are required, then it is advantageous to freeze the key row and/or column titles so that they remain in view as you scroll through the worksheet. For worksheet 6.1c and 6.2c the following steps would freeze the title:

1. Move the cell pointer to G34 – which is just to the right of the depth column and just below the title headings.

2. Enter/Worksheet Titles Both command.

As you scroll through the worksheet the titles and depth will remain in view. To remove titles use /WT Clear.

There are two main disadvantages to the /WT command. Firstly, you cannot move freely between the title area and the body of the worksheet and secondly you are confined to freezing rows and columns to the top and left of the worksheet respectively. The /Worksheet Window command provides greater flexibility in this regard. You can change windows with the (F6) function key.

Alternatively the windows and titles may be used in combination.

HINT: For editing a label in the title area without clearing the titles then use the GOTO key (F5) and the arrow keys to move into the titled area. To get out of the title, just press ENTER.

Ult Bearing Capacity (kN/m/m)		Ultimate Load (kN)		Total Allowable	No of Piles	Single Pile
Base q(b)	Shaft q(s)	Base Q(b)	Shaft Q(s)	Load (kN) Q (all)	required	Capacity
1309.1	12.0	99.0	19.8	46.2	1	1000
6000.0	21.0	453.8	54.5	187.6	1	1000
7200.0	26.0	544.5	97.4	246.4	1	1000
4800.0	25.5	363.0	139.4	201.0	1	1000
9200.0	29.6	695.8	188.3	353.6	1	1000
10800.0	33.7	816.8	243.8	424.2	1	1000
12800.0	38.0	968.0	306.5	509.8	1	1000
15200.0	42.8	1149.5	377.1	610.6	1	1000
17200.0	47.6	1300.8	455.5	702.5	1	1000
22000.0	53.8	1663.8	544.3	883.2	1	1000
22400.0	59.1	1694.0	641.8	934.3	1	1000
19200.0	62.2	1452.0	744.4	878.5	1	1000

Worksheet form 6.2. Data entry and output table. Filename: Pilegra.

```
                              DESIGN OF PILES
                              ***************

The program considers the following :
                        pile type (bored or driven)
                        soil type (cohesive or cohesionless)
       Notes :          variation of pile capacity with length and size
       --------
   – Meyerhof's S.P.T. correlations are used for piles in Cohesionless soils
   – Nordlund's adhesion factors are used for driven piles in Cohesive soils
   – Tomlinson/Skempton's adhesion factors used for bored piles in Cohesive soils
   – Geotechnical Models adopted (4) are used solely on the basis of Nordlund's 3 models ( +1)

Design Considerations :                              Design Conditions
---------------------.                               ---------------.------.


What type of pile is proposed – (D)riven or (B)ored ?          b
                        Pile size (shaft)                   950 mm
                        Pile size (Base)                    950 mm
        Pile Shape: (R)ound, (S)quare or (O)ctogonal ?        o
                        Estimated pile load                 800 kN
                   Design Structural Capacity              1000 kN
                   Reduced Level at Surface               18.42 metres


                   Factor of safety (Bearing)                3
                   Factor of safety (Shaft)                 1.5
                   Factor of safety (Overall)               2.5


Soil Type – (C)ohesive clays or (N)on–cohesive ?             c
Is the degree of consolidation of the clays                 f
(N)ormal, (L)ightly or (H)ighly Overconsolidated, or (F)issured ?
--------.
```

```
Design of Bored piles in fissured clays
-----------------------------.-------. ---------------.-----------------.
              Undrained  Adhesion   Ult Bearing capacity   Ultimate Load
Depth   Reduced  Shear      Coefficient (kN/m/m)                (kN)
(metres)  Level   Strength   of Bored    Base      Shaft      Base      Shaft
                  Cu (kPa)   Piles       q(b)      q(s)       Q(b)      Q(s)
-----------------------------.---------------------------------.--------.----------.
```

Depth (metres)	Reduced Level	Undrained Shear Strength Cu (kPa)	Adhesion Coefficient of Bored Piles	Ult Bearing capacity (kN/m/m) Base q(b)	Shaft q(s)	Ultimate Load (kN) Base Q(b)	Shaft Q(s)
0	18.42	0.00					
1.8	16.62	0	0.30	0.0	0.0	0.0	0.0
4.42	14	100	0.30	600.0	30.0	425.3	234.6
8	10.42	200	0.30	1200.0	60.0	850.6	875.7
8	10.42	450	0.30	2700.0	135.0	1913.8	875.7
9	9.42	450	0.30	2700.0	135.0	1913.8	1278.6
9.5	8.92	450	0.30	2700.0	135.0	1913.8	1480.0
10	8.42	450	0.30	2700.0	135.0	1913.8	1681.5
10.5	7.92	450	0.30	2700.0	135.0	1913.8	1882.9
11	7.42	450	0.30	2700.0	135.0	1913.8	2084.4
11.5	6.92	450	0.30	2700.0	135.0	1913.8	2285.8

```
-----------------------------.-----.  -----------------------.-----------.
Borehole 2
```

```
Geotechnical Model is considered one layered – Use Model 4
------------------------------------------------------------
Geotechnical Model    |        Model 1 : Sand/gravel over Stiff Clay
         1            |        Model 2 : Soft Clay over Stiff Clay
Sand/Gravel thickness |        Model 3 : Only Stiff Clay
       1.8    metres  |        Model 4 : One Layered Soil Strata
------------------------------------------------------------
```

Total Allowable Load (kN) Q (all)	No of Piles required	Pile Capacity	Percentage Carried by Base	Shaft	Total Pile	Check on embedment L/D for that layer	Ult Brg Cap Base Nc	q(b)
0.0	1	800	ERR	ERR	0.0%	1.9	7.6	0
264.0	1	800	64.5%	35.5%	33.0%	2.8	8.3	830
690.5	1	800	49.3%	50.7%	86.3%	3.8	8.8	1760
1115.8	2	800	68.6%	31.4%	139.5%	0.0	6.3	2827
1277.0	2	800	59.9%	40.1%	159.6%	1.1	9.0	4050
1357.5	2	800	56.4%	43.6%	169.7%	0.5	9.0	4050
1438.1	2	800	53.2%	46.8%	179.8%	0.5	9.0	4050
1518.7	2	800	50.4%	49.6%	189.8%	0.5	9.0	4050
1599.3	2	800	47.9%	52.1%	199.9%	0.5	9.0	4050
1679.9	3	800	45.6%	54.4%	210.0%	0.5	9.0	4050

Worksheet form 6.3. Data entry and output table. Filename: 2Lpile1.

	Brg Capacity factor	
	L/D	Nc

Geotechnical Model is considered one layered	0	6.3
+++++++++++++++++++++++++++++++++++	1	7.6
	2	8.3
Adhesion Factors for Driven Piles (after Nordlund)	3	8.8
	4	9.0

Geotechnical Model	|Model 1 : Sand/gravel over Stiff Clay
1	|Model 2 : Soft Clay over Stiff Clay
Sand/Gravel thickness	|Model 3 : Only Stiff Clay
1.8 metres	|Model 4 : Not applicable here

Undrained Shear Strength Cu (kPa)	Adhesion factor at depth embedded into stiff clay				
	L <= 10 B	15 B	20 B	30 B	L > 40 B
50	1	1	1	0.95	0.9
100	1	0.95	0.9	0.75	0.7
150	1	0.9	0.75	0.55	0.35
200	1	0.9	0.75	0.55	0.35

Table is for a Geotechnical Model of   Sand/gravel over Stiff Clay

Depth Increment	Depth (metres)	Undrained Shear Strength Cu (kPa)	Adhesion factor at depth embedded into stiff clay					
			9.50	14.25	19.00	28.50	38.00	Design Value
0	0	1E−05	0	0	0	0	0	0
1	1.8	0	0	0	0	0	0	0
2	4.42	100	1	0.95	0.9	0.75	0.7	1
3	8	200	1	0.9	0.75	0.55	0.35	1
4	8	450	1	0.9	0.75	0.55	0.35	1
5	9	450	1	0.9	0.75	0.55	0.35	1
6	9.5	450	1	0.9	0.75	0.55	0.35	1
7	10	450	1	0.9	0.75	0.55	0.35	1
8	10.5	450	1	0.9	0.75	0.55	0.35	1
9	11	450	1	0.9	0.75	0.55	0.35	1
10	11.5	450	1	0.9	0.75	0.55	0.35	1

Worksheet form 6.3 (continued).

CHAPTER 7

Shallow foundations

The design of shallow foundations involves the following considerations:

1. The bearing capacity of the soil must be adequate.

2. The resulting settlements for that foundation and building type must be within tolerable limits.

If any of the above is unacceptable or involves the use of an uneconomical size footing then deeper foundations will be required.

During the design process, the engineer is required to assume footing dimensions, to calculate the allowable bearing capacity. The calculated bearing capacity is then used to determine the foundation size. If the original assumption of footing dimensions is incorrect, then the process is repeated, until the footing dimension is suitable for the allowable bearing capacity.

This iteration process is common in conventional programming languages but is not as directly applicable in spreadsheet design. The facility however does exist to accomodate limited iteration effects as will be shown in the spreadsheet for bearing capacity analysis. Use of the BACKSOLVER add-in is also used to show a different approach to solving this problem. (The iteration facility to solve the problem is also available in the macro programming, but not discussed here).

Spreadsheet analysis is also used to examine settlements in sands allowing for the corrected 'N' values and stress distribution with depth. An enhanced pie chart is also illustrated. Reference should be made to Simons & Menzies (1979) or Das (1984) for the relevant theory.

7.1 BEARING CAPACITY

The general bearing capacity equation involves evaluation of:
- Bearing capacity factors (N_q, N_c, N_γ),
- Shape factors (S_c, S_q, S_γ),
- Depth factors (D_q, D_c, D_γ),
- Inclination factors (I_c, I_q, I_γ).

Besides the strength (cohesion c, friction angle ϕ, density γ) of the soil, the

Table 7.1. General bearing capacity equation.

$$q_u = c N_c S_c D_c I_c + q N_q S_q D_q I_q + \frac{1}{2} \gamma B N_\gamma S_\gamma D_\gamma I_\gamma$$

where: q_u = ultimate bearing capacity; c = cohesion; q = effective stress at the base of the footing = Q/B per unit length; γ = unit weight of soil removed; B = width of footing (= diameter for a circular footing).

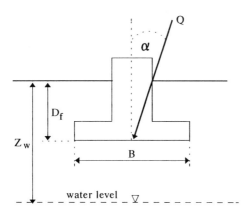

Figure 7.1. Design considerations for shallow footing.

allowable bearing capacity is therefore dependent on:
  – the footing size (B),
  – the depth of embedment of the footing (D_f),
  – the depth of the water table (Z_w),
  – the magnitude of the applied load (Q),
  – the inclination of the applied load (α).
These factors are illustrated in Figure 7.1.

The presence of the water table at or near the foundations will influence effective strength, overburden, and the density as reflected by the c, q and γ bearing capacity factors.

The net bearing capacity is the pressure per unit area in excess of the pressure caused by the surrounding soil at the founding level. It is very important to distinguish between the gross and net foundation pressures and loads.

Gross load = Total applied load
Net load = Net increase in loading = Total applied load – weight of soil removed.

Bearing capacity factors
The bearing capacity factor is based on the angle of friction of the soil with the relationships given in Table 7.2.

Table 7.2. Bearing capacity factors.

$$N_q = \tan^2 (45 + \frac{\phi}{2}) \exp^{(\pi \tan \phi)}$$

$$N_c = (N_q - 1) \cot \phi$$

$$N_\gamma = 2 * (N_q + 1) \tan \phi$$

Shape and depth factors

The shape and depth of the proposed footing will influence the bearing capacity as shown in Tables 7.3 and 7.4 respectively.

The shape factor equations show that the square and circular footings are a special case of the rectangular footing, when $B/L = 1$.

The depth factor is dependent on the angle of friction, the depth of the footing and footing size as well as the bearing capacity factor.

Table 7.3. Shape factors (S).

Rectangular	Square/circular
$S_q = 1 + \frac{B}{L} * \tan \phi$	$1 + \tan \phi$
$S_c = 1 + \frac{B}{L} * \frac{N_q}{N_c}$	$1 + \frac{N_q}{N_c}$
$S_\gamma = 1 - 0.4 \frac{B}{L}$	0.6

Table 7.4. Depth factors (D).

$$D_q = 1 + 2 \tan \phi \, (1 - \sin \phi)^2 \frac{D_f}{B} \quad \text{for} \frac{D_f}{B} \le 1$$

$$= 1 + 2 \tan \phi \, (1 - \sin \phi)^2 \tan^{-1} \left(\frac{D_f}{B}\right) \quad \text{for} \frac{D_f}{B} > 1$$

$$D_c = D_q - \frac{(1 - D_q)}{N_q * \tan \Phi} \quad \text{for } \Phi > 0: \frac{D_f}{B} \le 1$$

$$= 1 + 0.4 \frac{D_f}{B} \quad \text{for } \phi = 0: \frac{D_f}{B} \le 1$$

$$D_c = \frac{(1 - D_q)}{N_q * \tan \phi} \quad \text{for } \phi > 0: \frac{D_f}{B} > 1$$

$$= 1 + 0.4 \tan \phi^{-1} \left(\frac{D_f}{B}\right) \quad \text{for } \Phi = 0: \frac{D_f}{B} > 1$$

$$D_\gamma = 1 \quad \text{for: All values of } \Phi: \text{All values of} \frac{D_F}{B}$$

Table 7.5. Inclination factors (I).

$$I_q = I_c = \left(1 - \frac{\alpha}{90}\right)^2$$

$$I_\gamma = \left(1 - \frac{\alpha}{\phi}\right)^2$$

Inclination factors
The direction of the applied load will also influence the bearing capacity as given in Table 7.5.

7.2 SETTLEMENTS

For granular soils, settlement considerations rather than the shear strength of the soil will most likely govern the choice of allowable pressures for immediate effects.

Immediate settlements can be calculated using the relationships shown in Table 7.6, which approximates to the standard Terzaghi-Peck settlement chart and provides allowable pressures to give 25 mm settlements based on SPT 'N' values and width of footings for width of footing (B) and applied foundation pressure (q).

N is averaged over a depth equal to the width of the footing. It is therefore essential that the corrected N values are used in evaluation of settlements. The corrected N values are then used together with a reduced pressure with depth (based on elastic theory). The total settlement over the entire length is then summed over a depth of influence.

This is therefore the combination of three distinct aspects of soil mechanics theory.

1. Corrected N values   ... Existing condition
2. Stress distribution   ... Applied condition
3. Settlement theory   ... Resulting condition.

Different methods proposed by different researchers for the corrected Standard Penetration Test value produce some variation in the applied correction factor

Table 7.6. Settlement with footing size and SPT value.

Foundation size (metres)	B < 1.25	B > 1.25	Large rafts
Settlements (mm)	$\dfrac{1.9\,q}{N}$	$\dfrac{2.84\,q\,[B/(B+.33)]^2}{N}$	$\dfrac{2.84\,q}{N}$

Table 7.7. Correction factor for N value.

Effective overburden pressure (kPa)	Correction factor C_N
0	2
6	1.8
15	1.6
> 23.9	$0.77 \log (20/.0105\ P_0)$

where C_N = N (corrected value)/N (field value); $C_N = 1$ when the effective overburden pressure is 95.6 kPa.

especially in the near surface values. The method of Peck, Hanson & Thorburn (1974) is used herein. The N value correction factor is usually expressed in a graph but the values may be obtained from Table 7.7.

In addition, where the soil is a fine or silty sand below the water table then the SPT 'N'-Value is adjusted as:

N (corrected) = 15 + ½ (N-15)

The stress distribution effect as covered in Chapter 3 is then applied on to the existing condition. The resulting settlements over increments of depth can then be tabulated.

7.3 WORKSHEET DESIGN FOR SPREAD FOOTING ANALYSIS

PROBLEM: A square footing is to carry an applied load 150 kN. The footing is to be embedded at 0.7 m depth and of 0.3 m thickness reinforced concrete (unit weight of 24 kN/m³). The load is inclined at an angle of 20° to the vertical.
   The soil parameters are as follows:
   – Cohesion = 0 kPa,
   – Angle of friction = 30°,
   – Unit weight of soil = 18 kN/m³.
The water table is at 3 m below ground level.
   Calculate the width of footing required.
   Evaluate the expected settlements for a footing with the SPT profile as shown.

SOLUTION WORKSHEET: *Worksheet 7.1*
This shows the already completed worksheet for calculation of the footing size. Use this as a guide during the worksheet development.
   The following Lotus functions will be used:
   – Iteration facilities,
   – Circular calculations,
   – Descriptive feedback, rather than numerical output,
   – The use of screen forms,

- Use of logical operators #AND#, #OR# with the IF functions. This is in contrast with the embedded (NESTED @IF) functions used in former chapters.
- @Round function,
- @CHAR for character symbols,
- Windows,
- Backsolver add-in,
- Exploded pie charts,
- Combining worksheets.

Table 7.8. Worksheet logic.

Data input	Defined conditions
Soil properties	Footing size
Proposed footing geometry	
Loading conditions	
Groundwater conditions	
Compute	
Bearing capacity factors	Table 7.2
Shape factors	Table 7.3
Depth factors	Table 7.4
Inclination factors	Table 7.5
Compute	
Ultimate bearing capacity	Table 7.1
Allowable loads	
Calculate	*Iterate for optimum*
Footing size	Footing size

This example illustrates the form filling approach which may be used in the design process. The designer may then 'redesign' the footing by varying any one of the parameters on the form.

The calculation is mechanical once the fundamental equation is known and the geometry and soil parameters are defined. The equation is an iterative type whereby the size of the footing is defined initially and checked against the loading condition in order to define the size of footing required.

The flow of the calculations can be seen from the Table 7.8.

WORKSHEET 7.1: FOOTING DESIGN

Data input

The various elements of the worksheet have been 'boxed off' for ease of reading. The description part of the worksheet (columns H & I) are not shown above. The

```
      A    B       C        D       E      F    G    H       I      J
1    :::::::::::::  Spread Footing Design  :::::::::::::
2    :::::::::::::  :::::::::::::  :::::::::::::  ::::::::  :::::::::::::
3
4        Soil Properties              Value Units          Description
5     * ******* ******* *******   ***** *******  * ------  -------
6     *            Cohesion ?          0 kPa       *               |
7     *       Angle of Friction ?     30 °         * 0.5236  radians |
8     *        Bulk Unit Weight ?     18 kN/ cu m  *               |
9     *    Saturated Unit Weight ?    18 kN/ cu m  *               |
10    * ******* ******* *******   ***** *******  *               |
11                                                                |
12       Proposed Footing Geometry                                |
13    * ******* ******* *******   ***** *******  *               |
14    *        Width of Footing ?  1.329 metres    *               |
15    * Footing length, (D) or (S)?   s   metres    * (D)iamater:(S)quare |
16    *    Depth of Footing (Df) ?      0.7 metres    * Design size      |
17    *     Thickness of Footing ?      0.3 metres    * Required         |
18    * ******* ******* *******   ***** *******  *               |
19
```

Worksheet 7.1a. Display screen.

```
      A    B       C        D       E      F    G    H       I      J
19                                                                |
20       Loading Conditions                                       |
21    * ******* ******* *******   ***** *******  *               |
22    *    Depth of Groundwater ?      3 metres    * below surface level |
23    *    Applied column load ?     150 kN        * at top of footing   |
24    *  Load Inclination {alpha} ?   20 °         * 0.3491  radians     |
25    *   Unit Weight of concrete  ?  24 kN /cu m  *               |
26    *         Factor of safety ?     3           *               |
27    * ******* ******* *******   ***** *******                  |
28                                                                |
29       Calculated Quantities                                    |
30    * ******* ******* *******   ***** *******  *               |
31    * Vol of concrete in footing  0.530 cu metre *               |
32    * ==> Weight of Footing  =     12.7 kN        *               |
33    * Weight of equivalent soil     9.5 kN        *               |
34    * == > Net applied load  =    153.2 kN        *               |
35    * Initial overburden pressure 12.60 kPa       * at base of footing |
36    * ******* ******* *******   ***** *******  *               |
37                                                                |
```

Worksheet 7.1b. Display screen.

reader may refer to the appendix to see the input for these cell entries. The following should be noted.

– The soil properties are defined with the units: Cohesion – kPa; Angle of friction – degrees; Bulk unit weight – kN/m^3; Saturated unit weight – kN/m^3.

– Lotus has a limited character set, where a few symbols/characters may be used. The @Char(176) was used here to show the degrees symbol (°). The user

```
    A     B          C         D        E      F      G    H          I       J
37                                                                             |
38        Calculation Table of Bearing Capacity factors                       |
39  + +++++++  +++++++  +++++++  +++++  +++++++        ++++++  +++++++  +
40  +           Brg Capac  Geometry & Load factors       Ultimate Bearing   +
41  +           Factors     Shape  Depth  Inclination      Capacity (kPa)    +
42  + +++++++  +++++++  +++++++  +++++  +++++++        ++++++  +++++++  +
43  + q           18.40      1.58   1.15    0.60         254.87              +
44  + c           30.14      1.61   1.17    0.60           0.00              +
45  + {gamma}     22.40      0.60   1.00    0.11          17.86              +
46  +                                                                        +
47  + +++++++  +++++++  +++++++  +++++  +++++++        ++++++  +++++++  +
48        +       Total Ultimate Bearing capacity          272.74  kPa       +
49        +            Allowable Bearing capacity           90.91  kPa       +
50        +            Allowable Total gross load           160.6  kN        +
51        +            Allowable Total net load =           153.2  kN        +
52        + +++++++  +++++++  +++++  +++++++        ++++++  +++++++  +
53        = Width of Footing (B) required                    1.329  metres    =
54        = =======  =======  =====  =======        ======  ======= =
55
56
```

Worksheet 7.1c. Display screen.

should refer to the Lotus manual for the full list of the character set.

– The angle of friction is converted to radians since the Lotus computations for angles are carried out in radians. The conversion is placed in the same row for the angle of friction in order to maintain the sequence of calculations. This occurs in cell H7. It may have been placed elsewhere in the spreadsheet, but should be kept above the main tables of computations.

Cell (Format)	Display	Cell Entry Formula	Heading/Formula
H7	.5236	+E7/180*@PI	Radians = Angle{degrees}/180*π

Worksheet 7.1. Conversion to radians.

The proposed footing geometry is then defined. This forms the initial design assumptions and the pivot point for iteration.

– Before moving any further the user should set the recalculation to Manual mode.

`/ Worksheet Global Recalculation Manual`

The recalculation is set to manual (instead of the automatic default setting) and when the user enters or changes a value the spreadsheet does not recalculate (and waste user's time) until ALL the parameters are entered.

At cell E14 enter the cell reference +H53. This provides the link between the initial 'guesstimate' and the required footing dimension as calculated from the bearing capacity.

(If the recalculation mode had not been set to manual, and the link between H15 → E14 was established, then an error message would occur during subsequent formula input, since the spreadsheet is not complete in its development).

> REMEMBER: We must estimate the footing size to calculate the bearing capacity. However the design footing size is based on the bearing capacity. This requires an iterative loop in the design process. This will be explained further in the next section (Calculations).

A logic function is placed at cells H16 & H17 to provide an immediate response to the footing size selected. This acts as a prompt to repeat the iteration process. These cell entries (H16 & H17) will be explained in a later section.

The loading input is used to calculate the net load. The factor of safety is also defined.

Cell (Format)	Display	Cell Entry Formula	Heading/ Formula
E14 (F3)	1.329	+H53	Width of footing

Worksheet 7.1. Circular referencing.

Calculations

The rest of the worksheet is simply the input of the formulae from the Tables 7.1 to 7.5. Only the logical operators and circular referencing is explained below.

A circular reference is used at cell E14, where the calculated end result (Width Of Footing Required – H53) is used to input the footing width for initial assessment. This is part of the iteration process.

Cell (Format)	Display	Cell Entry Formula
C53	Width of Footing (B) required	@IF($E$15="d","Diameter of Footing required","Width of Footing (B) required")
H53		@IF(@ROUND(E34,1)=@ROUND(H51,1),E14, @IF(@ROUND(E34,1)>@ROUND(H51,1),E14+0.001,E14−0 .001))

Worksheet 7.1. Width of footing.

The allowable total net load is checked against the net applied load and one of the three comments will appear at H16:

- Footing size suitable, but check for economical size,
- Footing size unsuitable, adjust size,
- Design size required.

This answer is provided using the iteration features. The designer is provided with a rule driven assessment of the calculated size.

Cell (Format)	Display	Cell Entry Formula
H16	Design Size	@IF(@ROUND(E34,1)=@ROUND(H51,1),"Design size", @IF(@ROUND(E34,1)>@ROUND(H51,1),"Footing size unsuitable","Footing size suitable, but"))
H17	Required	@IF(@ROUND(E34,1)=@ROUND(H51,1),"Required", @IF(@ROUND(E34,1)>@ROUND(H51,1),"Adjust size","Check for economical size"))

Worksheet 7.1. Rounding for comparison.

The use of @Round (cell, No of decimal places) is used here because the allowable load is being checked against the applied load. In each formula cell (even with fixed format decimal places), the number is calculated to several decimal places more than is usually displayed. By limiting the number of decimal places in the @Round function then the degree of accuracy is limited.

If this was not done, the numbers will not equal each other i.e. 152.2344786...kN calculated for the allowable load will not equal to 152.2344785...kN for the applied load, despite the values being the same for all practical engineering purposes.

Therefore any practical comparison on the 'equivalence' between calculated numbers can be made using the @Round function to the required number of decimal places and *not* the formatting to a fixed number of decimal places.

Logical operators are used in contrast to the series of embedded IF functions in previous chapters. The format is:

#AND#
#OR#

These are used in the case of the shape factors to evaluate between the three cases of footing type (rectangular, square, circular).

Table 7.3, for instance, shows that the shape factors for square and circular footings use the same formula, while a rectangular footing requires a different formula to be applied.

Table 7.4 shows that the depth factor is dependent on the depth/width ratio (D_f/B) and the angle of friction ϕ. The logical operators are used to distinguish between the limiting conditions and to apply the appropriate depth factor equation within the @IF function.

The values are rounded as before to ensure a degree of accuracy to only 1 decimal place. The value of cell E14 is then either increased or decreased if the net applied load does not exactly equal the net allowable (to 1 decimal place).

The iteration procedure is accomplished by:

Cell (Format)	Display	Cell Entry Formula	Heading/ Formula
C43	18.40	@EXP(@PI*@TAN(H7)) *(@TAN(45/180*@PI+H7/2))^2	Bearing capacity factors : Table 5.2
C44	30.14	@IF(H7=0,5.14,(C43−1)/@TAN(H7))	
C45	22.45	2*(C43+1)*@TAN(H7)	
D43	1.58	@IF(E11="d"#OR#E11="s",1+@TAN(H7) ,1+(E14/E15)*@TAN(H7))	Shape factors : Table 7.3
D44	1.61	@IF(E15="s"#OR#E15="d",1+C43/C44, 1+(E14/E15)*(C43/C44))	
D45	0.60	@IF(E15="d"#OR#E15="s",0.6, 1−0.4*E14/E15)	
E43	1.15	@IF(E16/E14<=1,1+2*@TAN(H7) *(1−@SIN(H7))^2*E16/E14, 1+2*@TAN(H7)*(1−@SIN(H7))^2 *@ATAN(E16/E14))	Depth factors : Table 7.4
E44	1.17	@IF(E16/E14<=1#AND#E7=0, 1+0.4*E16/E14, @IF(E16/E14<=1#AND#E7>0, +E43−(1−E43)/C43/@TAN(H7), @IF(E7=0#AND#E16/E14>1, 1+0.4*@ATAN(E16/E14), +E43−(1−E43)/C43/@TAN(H7))))	
E45	1.00	1	
F43	0.60	(1−E24/90)^2	Inclination factors : Table 7.5
F44	0.60	(1−E24/90)^2	
F45	0.11	@IF(E7=0,1,(1−E24/E7)^2)	

Worksheet 7.1. Bearing capacity factors.

/ W G Recalculation Iteration {Set Counter to 50}

The larger the counter setting then the more iterations would occur. The counter is set to check the logic function of net applied load to the allowable net load for comparision. For a simple case as this, a convergence is expected within a few iterations. The {F9} RECALCULATION key may have to be used until the dimension at cell E14 equals the calculated dimension at cell H53. This depends on the closeness of the initial guesstimate to the true dimension.

Use the Window feature to simultaneously view the input E14 and the output H53. Place the cursor about row 18 and use the commands:

/ Worksheet Window Horizontal

Use {F6} WINDOW go to the lower window and move the cursor to row 53. Press {F6} to return to the upper window.

The effect of the input on the output can now be viewed simultaneously.

The recalculation was set to manual mode at the start of the worksheet. Once *all* values have been entered (unprotected cells) the recalculation is done by using the {F9} function key. For large spreadsheets and, especially during spreadsheet development, it is useful to use this facility of manual recalculation.

The use of the BACKSOLVER add-in can also be used to solve the above problem in lieu of the ITERATION facility or simply to backcalculate any input parameters to a given constraint. For example, if the footing width cannot exceed 1.2 metres (due to the interaction with another structure) then this constraint can be used to backcalculate the required embedment depth to achieve a suitable design. The following steps are required.

1. ALT + F10 to load an add-in into memory
2. Select BACKSOLVER
3. Assign ALT + F9 to this application
4. ALT + F9 to invoke BACKSOLVER when required
5. Formula – cell: H53 ...Width of footing
6. Value 1.2
7. Adjustable: E16 ...Depth of footing
8. Solve

The depth required in this case will backcalculate to 0.835 metres.

This add-in or SOLVER can be useful to backcalculate the variables which would be applicable to determine the sensitivity of a design to any given parameter and to generate 'what if' conditions. It is useful where we guesstimate the solution then refine our estimate. It optimises a solution and is considered a goal seeking program, where variables are adjusted to achieve the goal.

WORKSHEET 7.2: SETTLEMENTS

	A	B	C	D	E	F	G	H
1	Settlements in granular Soil							
2	-----							
5	Proposed Foundation type (R)aft or (F)ooting) ?					f	Footing	
6			Proposed Footing size B			1.50	metres	
7			Applied Foundation Pressure			100.00	kPa	
8			Unit weight of soil			20.00	kN / cu m	
9			Depth to water table			10.00	metres	
10	Is the soil below the water Fine or Silty sand ?					Y	(Y)es or (N)o	
11	+++++ +++++++++++ +++++++++ +++++++++ ++++++++ ++++++ +++++++ ++++++							
12		S.P.T.	Effective	Corrected N Value		Applied	SETTLEMENT mm	
13	Depth	N Value	Overburden	For	For	Stress	at depth	Total
14	metres	Uncorrected	Press. kPa	Overburden	Saturation	kPa	increment	at depth
15	+++++ +++++++++++ +++++++++ +++++++++ ++++++++ ++++++ +++++++ ++++++							
16	1.50	8.00	30.00	11	11	47.51	8.2	8.2
17	3.00	12.00	60.00	14	14	28.99	4.0	12.2
18	4.50	15.00	90.00	15	15	20.31	2.5	14.7
19	6.00	9.00	120.00	8	8	15.52	3.6	18.2
20	7.50	7.00	150.00	6	6	12.53	4.0	22.3
21	9.00	12.00	180.00	9	9	10.49	2.1	24.4
22	10.50	15.00	205.10	11	13	9.02	1.3	25.7
23	12.00	13.00	220.40	9	12	7.91	1.2	26.9
24	13.50	20.00	235.70	14	14	7.04	0.9	27.9
25	15.00	25.00	251.00	17	16	6.34	0.8	28.6
26	+++++ +++++++++++ +++++++++ +++++++++ ++++++++ ++++++ +++++++ ++++++							

Worksheet 7.2. Display screen.

The worksheet has been set up to allow an easy evaluation of each stage of the calculations. The field N values with depth is first input (columns A & B). The effective overburden with depth is then calculated (column C) and the corrected N values determined for overburden (column D) and saturation effects (column E).

The stress distribution variation with depth is checked for a particular size footing and loading conditions (column F). The settlements are then calculated for each depth increment using the formula in Table 7.6 and the resulting total settlement in columns G and H respectively.

This model may be further expanded by including total allowable settlement for a particular type of building or differential allowable settlement if the distance between loading points and soil variability is also taken into account. The above specifications of allowable settlements would be entered into the cells as IF. . . THEN rules so that an immediate assessment is made with each calculation.

The reader is left to follow this worksheet through, using the appendix listings as required, as no new spreadsheet facility is introduced in this worksheet.

The following Lotus functions will now be shown:
 - Pie chart graphing,
 - Exploded pie charts for emphasis.

Graph
A pie chart is used to illustrate the settlement variation with depth. To create a pie chart the X, A, and B ranges may be used for specifying the pie headings, slice value (and proportion), and enhancements/colour/shadings for the respective ranges.

The pie chart was created as follows:

```
/ Graph Type Pie
            X {A16.A25}
            A {G16.G25}
            B {I16.I25}

            Options Title {Graph headings}
```

This produces a pie chart showing the relative percentages of the total settlement with depth shown. In order to enhance the graph a logic function is created in column 1, rows 16 to 25.

Cell I16: `@IF(G16>0.1*@SUM($G$16..$G$25),1,`
`        @IF(G16<0.05*@SUM($G$16..$G$25),100,0))`

This checks for three possible conditions of the value of the incremental settlements:

1. Above 10% of the total settlement → shades the segment of the chart `{0.1*@SUM($G$16..$G$25),1}`.

2. Below 5% of the total settlements → explodes the segment for emphasis `{0.05*@SUM($G$16..$G$25),100}`.

3. All other values are non shaded and non emphasised.

This was based on the assumption that values above 10% of the total settlement are significant in the analysis, while values less than 5% are considered insignificant. Values between 5 and 10% would be significant only for special cases e.g. movement sensitive structures.

A value of 0 or 8 provides an unshaded segment of the pie chart while values between 1 and 7 provide various shadings or colours. A value of 100 added to any of the above will produce an 'exploded' segment.

Therefore condition 1 had a shading defined by the value 1. Condition 2 (the exploded segment) is defined by the value 100, which is unshaded. Condition 3 is defined by the value 0, and this produces no shading.

As this is purely for enhancement of the pie chart, the cell values are hidden.

The B range for the graphing facility is then used to reference the I16 to I25 cell range and so produce the enhanced pie chart shown.

SETTLEMENTS IN GRANULAR SOILS
at each depth increment

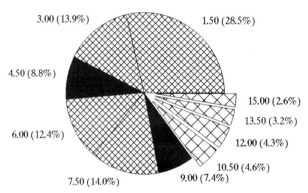

Graphsheet 7.1. Pie chart of the percentage settlements at each depth increment.

This worksheet can now be combined with the previous worksheet as follows:

/ W Erase
/ F Retrieve {Found}

{GOTO} Cell A60

/ File Combine Copy {Settle}

This file will now have to be linked with the main worksheet in terms of the allowable bearing pressure and footing dimensions. Therefore at cell F6 put +H59. Similarly for the footing dimensions and water table.

```
::::::::::::::::::::::  Spread Footing Design      ::::::::::::::::::

    Soil Properties                   Value Units           Description
    ::::::::::::::::::::::::::::::::::::::::::::::::::::::::::::::::::::::
    *               Cohesion ?           0 kPa          *
    *       Angle of Friction ?          30 °           *     0.5236 radians
    *           Bulk Unit Weight ?       18 kN/ cu m    *
    *   Saturated Unit Weight ?          18 kN/ cu m    *
    ::::::::::::::::::::::::::::::::::::::::::::::::::::::::::::::::::::::

    Proposed Footing Geometry
    ::::::::::::::::::::::::::::::::::::::::::::::::::::::::::::::::::::::
    *           Width of Footing ?       1.329 metres   *
    *Footing length, (D) or (S)?      s        metres   * (D)iamater:(S)quare
    *   Depth of Footing (Df) ?          0.7 metres     * Design size
    *   Thickness of Footing ?           0.3 metres     * Required
    ::::::::::::::::::::::::::::::::::::::::::::::::::::::::::::::::::::::

    Loading Conditions
    ::::::::::::::::::::::::::::::::::::::::::::::::::::::::::::::::::::::
    *   Depth of Groundwater ?           3 metres       * below surface level
    *     Applied column load ?          150 kN         * at top of footing
    * Load Inclination {alpha} ?         20 °           *    0.3491 radians
    * Unit Weight of concrete  ?         24 kN /cu m    *
    *           Factor of safety ?       3              *
    ::::::::::::::::::::::::::::::::::::::::::::::::::::::::::::::::::::::

    Calculated Quantities
    ::::::::::::::::::::::::::::::::::::::::::::::::::::::::::::::::::::::
    *Vol of concrete in footing          0.530 cu metre  *
    *==> Weight of Footing  =            12.7 kN         *
    *Weight of equivalent soil           9.5 kN          *
    *== > Net applied load  =            153.2 kN        *
    *Initial overburden pressure         12.60 kPa       * at base of footing
    ::::::::::::::::::::::::::::::::::::::::::::::::::::::::::::::::::::::

    Calculation Table of Bearing Capacity factors
    +++++++++++++++++++++++++++++++++++++++++++++++++++++++++++++++++++++
    +       Brg Capac    Geometry & Load factors        Ultimate Bearing   +
    +       Factors      Shape  Depth  Inclination      Capacity (kPa)     +
    +++++++++++++++++++++++++++++++++++++++++++++++++++++++++++++++++++++
    +q            18.40      1.58    1.15      0.60      254.87             +
    +c            30.14      1.61    1.17      0.60       0.00              +
    +{gamma}      22.40      0.60    1.00      0.11      17.86              +
    +                                                                      +
    +++++++++++++++++++++++++++++++++++++++++++++++++++++++++++++++++++++
    +     Total Ultimate Bearing capacity               272.74 kPa         +
    +            Allowable Bearing capacity             90.91 kPa          +
    +            Allowable Total gross load             160.6 kN           +
    +            Allowable Total net load =             153.2 kN           +
    + +++++++++++++++++++++++++++++++++++++++++++++++++++++++++++++++++++
    = Width of Footing (B) required                     1.329 metres      =
    = =================================================================
```

Worksheet form 7.1. Data entry and output table. Filename: Foundbrg.

Settlements in granular Soil

Proposed Foundation type (R)aft or (F)ooting ?					f	Footing	
Proposed Footing size B					1.50	metres	
Applied Foundation Pressure					100.00	kPa	
Unit weight of soil					20.00	kN / cu m	
Depth to water table					10.00	metres	
Is the soil below the water Fine or Silty sand ?					Y	(Y)es or (N)o	

+++

	S.P.T.	Effective	Corrected N Value		Applied	SETTLEMENT mm	
Depth	N Value	Overburden	For	For	Stress	at depth	Total
metres	Uncorrected	Press. kPa	Overburden	Saturation	kPa	increment	at depth

+++

Depth	S.P.T. N Value	Effective Overburden Press. kPa	Corrected For Overburden	For Saturation	Applied Stress kPa	at depth increment	Total at depth
1.50	8.00	30.00	11	11	47.51	8.2	8.2
3.00	12.00	60.00	14	14	28.99	4.0	12.2
4.50	15.00	90.00	15	15	20.31	2.5	14.7
6.00	9.00	120.00	8	8	15.52	3.6	18.2
7.50	7.00	150.00	6	6	12.53	4.0	22.3
9.00	12.00	180.00	9	9	10.49	2.1	24.4
10.50	15.00	205.10	11	13	9.02	1.3	25.7
12.00	13.00	220.40	9	12	7.91	1.2	26.9
13.50	20.00	235.70	14	14	7.04	0.9	27.9
15.00	25.00	251.00	17	16	6.34	0.8	28.6

+++

Worksheet form 7.2. Data entry and output table. Filename: Foundset.

CHAPTER 8

Ground improvement

Ground improvement by consolidation involves the evaluation of embankment settlements with time, and the effects of varying surcharge and vertical drains on the rate of consolidation. In addition, the incorporation of stress analysis, bearing capacity theory and construction procedures provide a more integrated design approach, but often involves lengthy and repetitive calculations. Design charts applied to this type of problem are developed herein to show the merits of alternative designs.

Emphasis in this chapter will be on transferring the numerical results into a graphical form with support comments on the graph, and building on the spreadsheet facilities already explained in previous chapters.

8.1 CONSOLIDATION THEORY

8.1.1 *Soil types*

Consolidation is the gradual change of volume which occurs in a low permeability soil during drainage of excess pore water. The way in which the soil responds to loading and its consolidation effects will be determined in part by its stress history.

A normally consolidated soil has not experienced any stress greater than the present stress state, while an overconsolidated soil has experienced higher stresses than presently exists. The higher stresses may be due to erosion of overburden soil, water level fluctuation, etc.

The existing overburden pressure (P_o) and the maximum past overburden pressure (P_c) define the stress history of the soil.

Consolidation settlement occurs as a gradual process and the engineer is generally interested in the magnitude of the settlement, the time frame under which it occurs as well as methods of accelerating the settlement process.

Reference can be made to Das (1984) or Craig (1983) for the relevant theory.

Table 8.1. Definition of soil types.

Stress history	Soil type
$P_o = P_c$	Normally consolidated
$P_o < P_c$	Lightly consolidated
$P_o \ll P_c$	Heavily overconsolidated

8.1.2 *Magnitude of settlement*

For a thickness of compressible layer (h – metres) and an increased loading of δP (kPa), then the magnitude of the settlements will be based on the applied load and the stress history of the soil.

For thin soil layers and/or wide embankments such that the stress level is approximately constant the settlement (δh) may be calculated from Table 8.2 where m_v is the coefficient of volume compressibility (m^2/kN).

The coefficient of compressibility, m_v, is not a fundamental parameter and shows large variations with stress level and should be used only for thin layers of clays. It should be used only in materials which are not sensitive to stress variations, or where the m_v stress changes are also included in the analysis.

A detailed analysis of consolidation settlement would involve the equations of Table 8.3.

Table 8.2. Consolidation settlement.

$$\delta h = m_v * h * \delta P$$

Table 8.3. Settlement relationships for different soil types.

Soil type	Settlement δh
Normally consolidated ($P_o = P_c$)	$\dfrac{C_c * h}{1 + e_o} * \log \dfrac{(P_o + \delta P_{av})}{P_o}$
Overconsolidated ($P_o + \delta P_{av} < P_c$)	$\dfrac{C_s * h}{1 + e_o} * \log \dfrac{(P_o + \delta P_{av})}{P_o}$
Overconsolidated ($P_o + \delta P_{av} > P_c$)	$\dfrac{C_s * h}{1 + e_o} * \log \dfrac{P_c}{P_o} + \dfrac{C_c * h}{1 + e_o} * \log \dfrac{(P_o + \delta P_{av})}{P_c}$

where: C_c = compression index (slope of the virgin compression curve for a void ratio versus log pressure plot); C_s = Swelling index (slope of rebound curve). The swelling index is, in most cases 5 to 10% the compression index value; e_o = initial void ratio; P_o = overburden pressure (kPa); P_c = preconsolidation pressure (kPa); δP_{av} = average increase of pressure (kPa) on the clay layer.

In addition, all of the above relationships are based on consolidation theory, where zero lateral strain occurs in the Odeometer test and the pore pressure parameter $A = 1$. A is defined as the ratio of change of pore pressure (δu) with vertical loading increase (δP):

$$A = \frac{\partial u_1}{\partial P_1}$$

The value of the parameter A at failure will vary with the type of soil.

8.1.3 *Rate of settlement*

The time for consolidation is based on the degree of consolidation which is determined from a non dimensional time factor constant. This can be established from standard laboratory consolidation tests and calculated from the relationship shown in Table 8.4.

The time factor constant (T_v) is a function of the time (t), the thickness (h) of the layer, drainage path, and the coefficient of consolidation (c_v) of the soil. For two way drainage, the length of the drainage path is half the thickness.

The coefficient of consolidation is dependent on the permeability (k), volume compressibility (m_v) of the soil and the unit weight of water (γ_w).

The time factor may then be used to establish the degree of consolidation as shown in Table 8.5.

Table 8.4. Time factor value.

Time factor	Coefficient of consolidation
$T_v = \dfrac{C_v t}{h^2}$	$c_v = \dfrac{k}{m_v \gamma_w}$

Table 8.5. Time factor relationship.

Time factor T_v	Degree of consolidation (U)
$\dfrac{\pi}{4}\left(\dfrac{U\%}{100}\right)^2$	for $U = 0 - 60\%$
$1.781 - 0.933 \log (100 - U\%)$	for $U > 60\%$

8.2 SOIL IMPROVEMENT

Soil improvement may cover improvement of strength or minimisation of post construction settlements to tolerable limits. Only the latter is covered herein.

Soil improvement may occur by surcharging, installation of vertical drainage (sand/wick drains), lime columns, dynamic compaction, etc.

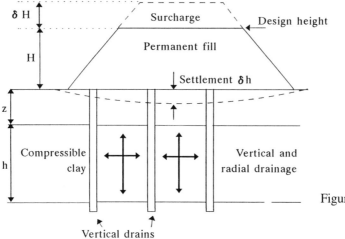

Figure 8.1. Site conditions.

Table 8.6. Design consideration.

Vertical drains	Surcharge	Construction
Sand/wick drains	Height possible	Placement time of surcharge
Triangular or square pattern arrangement	Unit weight of fill	Placement time of vertical drains
Spacing	Berm required for stability	Required degree of consolidation
Equivalent radius (smear effect)	Stability (slope and bearing)	Time for removal of surcharge

Typical analysis uses a combination of vertical drainage and surcharging to induce the settlements as shown in Figure 8.1.

Design considerations would involve evaluation of the factors given in Table 8.6.

Surcharge
Post construction primary consolidation settlements may be eliminated or considerably reduced by precompression of the soil. This involves placing an equivalent weight of soil as the expected loads. Depending on the scheduling of the project, the precompression process may be accelerated by a surcharge weight, in addition to the expected loads.

The placing of surcharge fill will however be constrained by:
– the stability characteristics, requiring a possible staged loading and/or stabilising berms,
 – the marginal effect (minimal gain) above a certain height,
 – availability of the fill material space available,
 – the time taken for placement and removal of the fill, which may offset the actual gain in consolidation time. (The fill should not be placed too rapidly as this may in turn produce instability due to excess pore pressure build up).

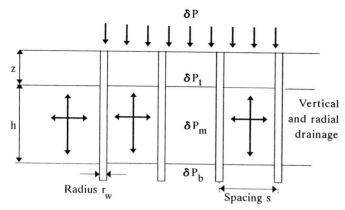

Drain pattern : square or triangular

Figure 8.2. Vertical drainage.

Vertical drains

Sand or wick drains can be used to speed the consolidation process by providing a radial consolidation path as shown in Figure 8.2. Because an alluvial soil deposit is often stratified or contains horizontal silt or sand lenses, the vertically placed drain connects these thin lenses and reduces the length of the drainage path.

The number and size of the vertical drains combined with the amount of surcharge will determine the degree and time of consolidation. It is readily evident that in order to optimise the use of time, resources and money, a wide range of options should be supplied to the designer and project manager to select the optimum alternative.

The placing of wick drains was found to be faster and cheaper than sand drains and is generally the preferred option. The presence of the sand drains would, however, provide a strength increase above that attributed solely to consolidation effects i.e. the columnar effect.

Design of vertical drains

The vertical drains would produce horizontal and vertical drainage. Analogous parameters for the two types of drainage are shown in Table 8.7.

The radial time factor constant (T_r) is a function of the time (t), the spacing of the drains (S), drainage path, and the coefficient of consolidation (C_r) of the soil. For two way drainage, the length of the drainage path is half the drain spacing.

The radial drainage mechanism is controlled by the drain spacing (S) and the

Table 8.7. Drainage parameters.

Drainage	Vertical	Horizontal (radial)
Time factor constant	T_v	T_r
Coefficient of consolidation	c_v	c_r
Drainage path (two way)	$h/2$	$S/2$

pattern arrangement (square or triangular) as well as the degree of soil disturbance expected on installation of the drains (a smear zone).

Allowance should be made for the effect of the smear zone. However the use of a number for a smear zone represents a mathematical convenience rather than a reality since the smear effects will not be uniform between the drains, with the smear effect more closely felt near the edge of the drain.

NOTE: A multiplication factor of 1.5 is used to account for the smear zone in this example.

Due to the effect of layering during deposition, the soil permeability in the horizontal direction is usually over 10 times the permeability in the vertical direction. The volume of compressibility is however considerably less in the horizontal than the vertical, but the relative difference is not as significant as the permeability. Therefore, the net effect of the above produces a greater horizontal consolidation than vertical consolidation.

NOTE: The example given in this chapter calculates the horizontal consolidation coefficient by using a factor of two applied to the vertical consolidation value.

Conservative designs are produced by equating the radial with the vertical coefficient of consolidation, but this can result in a severe over estimation of times for consolidation and the effects of radial drainage. The literature is not specific on the relationship except to say that there is greater horizontal than vertical consolidation effects. Often the key factor providing a good estimation of the times for soil improvement (comparing actual versus predicted), would be the initial estimation of the coefficient of radial consolidation.

Table 8.8. Effect of drain pattern.

Square	Triangular
$d_e = .564 * S$	$d_e = .525 * S$

Table 8.9. Average degree of radial consolidation (U_r).

$$U_r = 1 - \exp\left(\frac{-8 T_r}{m}\right)$$

where

$$m = \left(\frac{n^2}{n^2 - 1}\right) \ln(n) - \frac{3 n^2 - 1}{4n^2}$$

$$n = \frac{d_e}{2r_w}$$

The pattern spacing (S) would vary to produce the equivalent spacing (d_e) or equivalent radius (R) as shown in Table 8.8.

The vertical drainage will produce a radial consolidation effect and is calculated from Table 8.9.

The total degree of consolidation is then calculated as the combined effect of the vertical and radial consolidation as given in Table 8.10.

The induced settlements (S_p) due to preloading and/or vertical drains is determined from Table 8.11.

Table 8.10. Average degree of consolidation.

$$(1 - U) = 1 - (1 - U_v)(1 - U_r)$$

Table 8.11. Induced settlements.

$$S_p = U * (S_p + S_f)$$

where: S_p = settlement from preloading fill; S_f = settlement due to design fill height.

8.3 STRESS INCREASE UNDER AN EMBANKMENT

For thick layers or where a narrow embankment is to be built the effect of stress distribution with depth has to be considered. As the layer will experience different magnitudes of stress increase this has to be considered in the analysis. The increase in pressure (δP) decreases with increasing depth. The average increase of pressure (δP_{av}) can be approximated by Table 8.12.

The embankment will increase the stress in the natural soil, but this stress increase is expected to decrease with depth. The vertical stress increase (δP) under an embankment is given by Table 8.13.

Table 8.12. Average increase of pressure.

$$\delta P_{av} = \frac{(\delta P_t + 4 * \delta P_m + \delta P_b)}{6}$$

where: δP_t = stress increase at top of layer; δP_m = stress increase at middle of layer; δP_b = stress increase at bottom of layer.

Table 8.13. Stress increase under an embankment.

$$\delta P = \frac{q}{\pi} * \left[\frac{(B_1 + B_2)}{B_2}(\alpha_1 + \alpha_2) - \frac{B_1}{B_2} * \alpha_2 \right]$$

where: $q = \gamma H$; γ = unit weight of the embankment soil; H = height of the embankment; $\alpha_1 = \tan^{-1}[(B_1 + B_2)/z] - \tan^{-1}(B_1/z)$; $\alpha_2 = \tan^{-1}(B_1/z)$

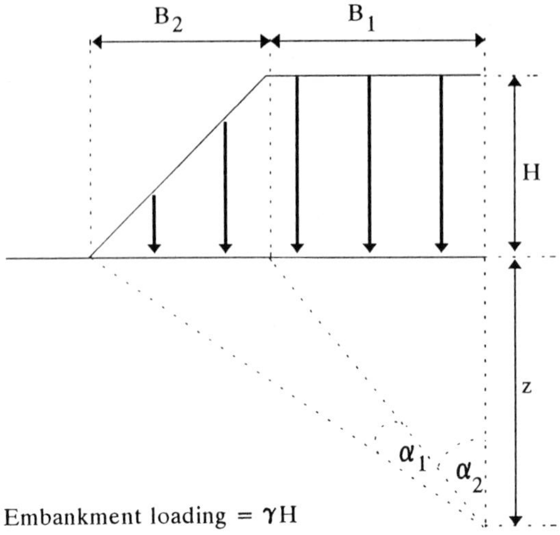

Embankment loading = γH

Figure 8.3. Loading from an embankment.

The geometric definition for a two dimensional analysis of the loading condition is shown in Figure 8.3. The analysis must be completed for the other side of the embankment. Therefore for a symmetrical embankment the stress below the middle would be calculated by using Table 8.13 and multiplying by 2.

8.4 BASE FAILURE

The short term stability of the fill may be checked for bearing capacity failure (Table 8.14).
   The result is compared with the design heights which may then dictate the construction procedures to be adopted. If the final design height is above that of the capacity of the soils then a staged loading will be required and/or the use of stabilising berms.

Table 8.14. Stability of fill.

$$H_{fill} = \frac{(2 + \pi) C_u}{\gamma \; FS}$$

where: H_{fill} = height of fill; C_u = undrained shear strength; γ = unit weight; FS = factor of safety (bearing).

8.5 SPREADSHEET DESIGN: CONSOLIDATION SETTLEMENT

PROBLEM: A site is to be developed for a canal industrial estate with material won from the canal excavation to be used to raise the level of the site. As this site is situated on the flood plains the final R.L. is required to be at 2.5 metres above the existing level with a 18 kN/m³ unit weight of compacted fill material.

This site had been inundated with water in the geologic past and deep alluvium was intersected during the site investigation. Marine clays of 5 m in thickness was met between sand layers. The clay was found to have the following properties:
- Coefficient of compressibility m_v: .001 m²/kN,
- Coefficient of vertical consolidation c_v: 5 m²/yr,
- Coefficient of radial consolidation $c_r = 2 * c_v = 10$ m²/yr.

In order to fully assess the construction options, the project manager requires information on the times for consolidation using various levels of surcharge with and without vertical drains.

Using a 3 m drain spacing of 150 mm diameter and arranged in a square pattern.

SOLUTION WORKSHEET: *Worksheet 8.1a, b, c*
Use this as a guide during the development.
Lotus functions:
- Use of an integrated spreadsheet,
- Blocking data area and output tables,
- Graphing facilities,
- Windows.

Two spreadsheets will now be used for the analysis of soil improvement. Spreadsheet 1 is a less detailed analysis, but covers the basic elements of any ground improvement. Spreadsheet 2 builds on that worksheet using the design elements of Section 8.3 and Table 8.3.

Two alternative spreadsheets may have been developed for evaluating consolidation settlements.

– Using a specified vertical drain design the settlements with varying surcharge are evaluated for a wide area of fill i.e. the effect of pressure distribution and base stability is not considered.

– Using a specified surcharge the settlements with varying vertical drain designs are evaluated for a specific embankment size (an access road to the main fill area).

Spreadsheet 8.1 uses the former and is developed for a wide embankment with a thin layer of compressible clay below. The stress with depth is assumed constant (extensive fill area) and settlements are calculated using the coefficient of volume compressibility only as given in Table 8.2.

Spreadsheet 8.2 also uses a specified vertical drain design for evaluating settlements with varying surcharge, but incorporates the effects of stress distribution and the possibility of bearing failure in the analysis. Spreadsheet 8.2 uses the relationships of Table 8.3.

This chapter draws on the Lotus techniques covered to this point to produce an integrated spreadsheet which spans various areas of the design process.

The input data required are:
- the soil and surcharge properties,
- geometrical arrangement of the vertical drains,
- time frames for design.

The computed table of values must first evaluate the degree of consolidation in the radial and vertical directions due to the presence of the vertical drains. Subsequently the combined effect is calculated as well as the variation of settlements with time for varying fill heights.

WORKSHEET 8.1: CONSOLIDATION SETTLEMENT FOR A VARYING FILL HEIGHT WITH A SPECIFIED VERTICAL DRAIN DESIGN

	A	B	C	D	E	F
1	CONSOLIDATION WITH TIME, FILL HEIGHT & VERTICAL DRAINS					
2	---------------- ----------- -------- -------- --------					
3		JOB TITLE :		Coomera		
4						
5		SOIL AND SURCHARGE PROPERTIES				
6	++++++++++++++++++ +++++++++++++ +++++++++ +++++++++ +++++++++ +					
7	Coefficient of compressibility Mv			0.001	sq m /kN	+
8	Coeff. of Consolidation (Vertical) Cv			5	sq m /yr	+
9	Coeff. of Consolidation (Radial) Cvr			10	sq m /yr	+
10	Thickness of compressible clay layer H			5	metres	+
11	Unit weight of placed fill			18	kN/cu m	+
12	Design Height of Fill			2.5	metres	+
13	Increment Fill Height			1	metres	+
14						+
15		PROPERTIES OF VERTICAL DRAINS				+
16	++++++++++++++++++ +++++++++++++ +++++++++ +++++++++ +++++++++ +					
17		Drain Radius rd		0.15	metres	+
18		Spacing of Drains S		3	metres	+
19	(S)quare or (T)riangular pattern		s			+
20		==> Equivalent radius (R) =			1.692	+
21	Factor to account for smear :		1.5 * R		2.538	+
22		========>		n =	8.46	+
23		========>		m =	1.419100	+

Worksheet 8.1a. Display screen.

	A	B	C	D	E	F	G	H
25			TIME FRAMES			+		
26	+++++++++++++++ +++++++++++ ++++++++ ++++++++ ++++++++ +							
27		Start of time t0		0	weeks	+		
28		End of time t1		40	weeks	+		
29	+++++++++++++++ +++++++++++ ++++++++ ++++++++ ++++++++ ++++++++ ++++++ +							
30		Time (weeks)	0	10	20	30	40	+
31		Time (years)	0.00	0.19	0.38	0.58	0.77	+
32	Radial Time factor Tr		0.00	0.30	0.60	0.90	1.19	+
33	Deg of radial consol Ur (%)		0.0	81.4	96.5	99.4	99.9	+
34	Vertical Time factor Tv		0.00	0.15	0.31	0.46	0.62	+
35	Deg of vert consol Uv (%)		0.0	44.3	62.1	74.0	82.2	+
36	Deg of Consol (rad+vert)%		0.0	89.6	98.7	99.8	100.0	+
37	+++++++++++++++ +++++++++++ ++++++++ ++++++++ ++++++++ ++++++++ ++++++ +							
38	Height of			Settlements (mm) expected				+
39	Fill (metres)	Total		with vertical drains				+
40	************** ********** ******* ******* ******* ******* ****** +							
41	2.5	225.0	0.0	201.7	222.1	224.6	225.0	+
42	3.5	315.0	0.0	282.4	310.9	314.5	314.9	+
43	4.5	405.0	0.0	363.1	399.7	404.3	404.9	+
44	5.5	495.0	0.0	443.7	488.5	494.2	494.9	+
45	6.5	585.0	0.0	524.4	577.3	584.0	584.9	+
46	7.5	675.0	0.0	605.1	666.2	673.9	674.9	+
47	************** ********** ******* ******* ******* ******* ****** +							

Worksheet 8.1b. Display screen.

	A	B	C	D	E	F	G	H
47	**********	*********	******	******	***********	******	*****	+
48								
49	\\\\\\\\\\\\\\\\\\	\\\\\\\\\\\\\\\\	\\\\\\\\\\\	\\\\\\\\\\\	\\\\\\\\\\\\\\\\\\\	\\\\\\\\\\	\\\\\\\\	\
50	Graph Aids to help illustrate the results							\
51	Design Settlement line				for fill height of		2.5 metres	\
52								\
53	Text illustrating line		675	675	675	675	675	\
54			650	650	650	650	650	\
55								\
56			Drain Radius (mm		150			\
57			Drain Spacing (m)		3			\
58			Square pattern					\
59								\
60	\\\\\\\\\\\\\\\\\\	\\\\\\\\\\\\\\\\	\\\\\\\\\\\	\\\\\\\\\\\	\\\\\\\\\\\\\\\\\\\	\\\\\\\\\\	\\\\\\\\	\
61								
62								

Worksheet 8.1c. Display screen.

Worksheet 8.1c is used only as an aid for graphing the results. This will be explained at the end of the section.

Worksheet 8.1 shows the display screen with the resulting Graphsheet 8.1. As this chapter is mainly concerned with the graphical output the reader should go through the calculation procedure in the spreadsheet using the worksheet shown as a guide with the appendix listings for reference.

Table 8.15 shows the calculation procedure adopted.

The graphical output is then developed, and this will be the main area of emphasis for this chapter.

Table 8.15. Soil improvement design procedure.

Input	The soil and surcharge properties
	Geometrical arrangement of the vertical drains
	time frames for design
Compute	Radial time factor, degree of radial consolidation
	Vertical time factor, degree of vertical consolidation
	Total consolidation, settlement with surcharge (fill height)
Evaluate	Effect of varying surcharge height, settlement with time

Input screen
The input screen deals mainly with text and numeric entries. The only area of calculation occurs where the effect of the drain pattern (Table 8.8) and size are used to calculate equivalent radius and, after allowing for smear effects, are used to calculate the *m* and *n* values (Table 8.9). This occurs at rows 20 to 23 in the worksheet.

Computation screen
These calculations are based on Tables 8.4, 8.5, 8.7, 8.9 and 8.10 for the time

factors and radial consolidations for vertical drainage, radial drainage and the combined effects respectively.

No new functions are introduced here, and the reader should be familiar with these input functions. The exponent function, logarithmic function and the IF functions are used.

The degree of consolidation at a specific time is calculated. The time interval is divided into 4 equal intervals in this analysis.

Evaluation screen

The coefficient of volume compressibility is used to calculate the total settlements (Table 8.2). The induced settlements for any given time (and degree of consolidation) is then calculated using the relationships in Tables 8.10 and 8.11.

The total settlement (mm) is the product of:

Stress increase * Thickness of layer * Coefficient of volume compressibility * 1000

and

Stress increase δP = Height of fill * Unit weight of fill

The settlements expected with time can then be evaluated for a specified vertical drain arrangement.

The rest of the worksheet deals with the graphing of the tabulated results. The graphical output has been structured to provide an automatic result i.e. without any user interface. Graphsheet 8.1 shows the graphical output and some detailed explanation will now follow.

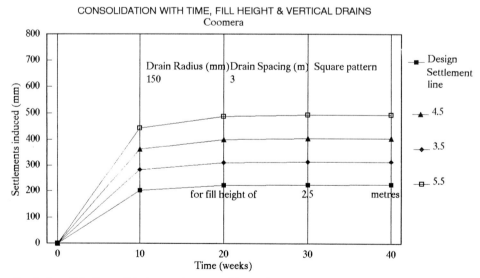

Graphsheet 8.1. Consolidation settlement with surcharge and vertical drains.

Graphing

A plot of induced settlements (mm) versus time (weeks) and for varying surcharge heights is shown in Graphsheet 8.1. The lowest line is the design fill settlement which will determine the minimum required induced settlements resulting from the combination of vertical drains and surcharge.

Cell referencing has been used extensively to automate the graph production and provide as much information given the wide variables to be considered.

Worksheet 8.1c showed the graph aids display screen and this is used as the basis for the additional graph display information.

The *Main Title* of the graph is referenced to the cell A1. The *Secondary Title* is the job title description, which is the input data at cell D3.

```
/ Graph Option Titles
            First                                     \A1
            Second                                    \D3
```

The lines are increments of fill heights as defined within the spreadsheet. The surcharge heights are labelled within the legend commands as follows:

```
/ Graph Options Legend
         A                          Design settlement line
```

Alternatively this may be typed into a cell and referenced by the backslash command and the cell reference: `\A51`.

The other legends are defined, such as

```
B: \A42 {the cell value of 3.5 is shown in this example}
C: \A43
D: \A44
```

In this way the numeric values contained in cells A41, A42, A43 are displayed in the legend. The cell values are in turn dependent on the defined fill heights and increments.

In order to create other explanation texts within the graphs, the data label function is used together with lines E and F. Up to this point the variable lines are referenced by A, B, C and D.

Logic functions are used to define whether vertical drains have been used and the geometry of the arrangement – if used. This is illustrated in the attached sub-spreadsheet which is an addendum to the main calculating body of the worksheet. This was input between rows 48 to 60 (under the heading *Graph aids*), and has been used solely to enhance the graph reporting function.

Graph aids

The actual cell entries are shown in the cell entry table.

The label on the *Design Settlement* line and the *Drain layout* description are created by using dummy lines as follows:

Graph Aids to help illustrate the results					
Design Settlement line			for fill height of		2.5
Text illustrating line		675	675	675	675
		650	650	650	650
		Drain Radius (mm)		150	
		Drain Spacing (m)		3	
		Square pattern			

Worksheet 8.1. Graph aids.

– The dummy line first calculates values to draw an imaginary line at the top of the graph.

– The format of the dummy line is adjusted by having *neither* the symbols nor the lines shown so the user is unaware of any referencing.

– Text is then incorporated by the use of the `Data-Labels` commands to

Cell	Display	Cell Entry Formula	Heading /Formula
A51	Design Settlement line	'Design Settlement line	Graph Aids to help illustrate the results
E51	for fill height of	'for fill height of	
F51	2.5	+D12	
G51	metres	'metres	
A53	Text illustrating line	'Text illustrating line	
C53	675	+$B$46	
D53	675	+$B$46	
E53	675	+$B$46	
F53	675	+$B$46	
G53	675	+$B$46	
C54	650	+C53–25	
D54	650	+D53–25	
E54	650	+E53–25	
F54	650	+F53–25	
G54	650	+G53–25	
C56	Drain Radius (mm)	'Drain Radius (mm)	
E56	150	@IF(D18>1000, "No vertical Drains",+D17*1000)	
C57	'Drain Spacing (m)	'Drain Spacing (m)	
E57	3	@IF(+D18>100, "No vertical Drains",D18)	
C58	Square pattern	@IF(+D19="t", "Triangular pattern","Square pattern")	

Worksheet 8.1. Formula entry for graphical output.

reference the created pseudo points. These then appear as text within the graph and provide much needed clarification for automatic reporting.

The logic functions for the text display on the graph should be noted.

The resulting graph then becomes fully interactive with the spreadsheet, and the need to constantly redefine the graph parameters and titles has been removed by this automation process.

WORKSHEET 8.2: CONSOLIDATION SETTLEMENT WITH A SPECIFIED VERTICAL DRAIN DESIGN AND WITH THE EFFECTS OF STRESS DISTRIBUTION

PROBLEM: An access road is to be built over a soft clay area as described in Problem 1. As this site is situated on the flood plains the final R.L. height of the road embankment is required to be at 2.5 m above the existing level with a 18 kn/m³ unit weight of compacted fill material. The embankment is 5 m wide at the top and with a side slope of 1 vertical to 2 horizontal.

Marine clays of 5 m in thickness were intersected with a surficial sand layer of 0.1 m just above the marine clay. The unit weight of the sand is 17 kN/m³. The clay was found to have the following properties:
- Preconsolidation pressure P_c: 25 kPa,
- Compression index C_c: 0.75,
- Initial void radio e_o: 0.6,
- Coefficient of compressibility m_v: .001 m²/kN,
- Coefficient of vertical consolidation c_v: 5 m²/yr,
- Coefficient of radial consolidation c_r: 10 m²/yr,
- Unit weight of clay γ: 15 kN/m³,
- Undrained cohesion c_u: 50 kPa.

The water table was intersected at 3 m depth.

In order to fully assess the construction options, the project manager requires information on the times for consolidation using various levels of surcharge with and without vertical drains. The stability of the fill also needs to be considered.

Use a 3 m sand drain spacing of 150 mm diameter and arranged in a square pattern.

SOLUTION WORKSHEET: *Worksheet 8.2a, b, c, d, e, f, g, h, i, j*
Use this as guide during the development.

Spreadsheet 8.1 is further enhanced to check the stability of the fill. Spreadsheet 8.2 examines the following:
- Settlement with time,
- Ground improvement,
- Surcharge and vertical (sand or wick) drains,
- Bearing capacity failure,
- Construction procedures outlined,

Table 8.16. Worksheet elements for soil improvement design.

Data input	Job description
	Embankment and surcharge properties
	Vertical drain design
	Time frames for design
	Soil profile
	Soil properties of compressible layer
Compute	Design considerations
	Consolidation parameters
	Influence factors with depth
	Settlement with time
Evaluate	Effect of varying surcharge height
	Settlement with time

- Effect of load distribution with depth,
- Effect of shape of embankment.

Each of the above can easily represent a distinct phase of the analysis process. However, the ease of spreadsheet development facilitates the integration into one analysis, with the flexibility to quickly examine changing conditions as required.

The worksheet elements are shown in Table 8.16. This worksheet includes the key factors covered in Spreadsheet 8.1, but is extended to cover other areas of analysis. The comments and intermediate calculations are placed under a 'Notes' heading to the right of the main data input screen and aids in the flow of the computed results.

Some comments and all graphing legends and headings are rule driven.

A graph aids box is included at the end of the worksheet and provides the referencing for the automatic graphing of the results. This will be used in the evaluation stage of any analysis.

This worksheet is therefore an integrated design of several aspects of soil mechanics and builds on the basics of Spreadsheet 8.1.

Specific features explained are:
- 'Expert' assessment of the parameters required if sand or wick drains,
- Computation of overburden pressure and comparison with allowable bearing pressure (from undrained cohesion value).
- Computation of stress increase based on the embankment geometry,
- Computation of settlements based on the type of soil (deduced from the preconsolidation pressure, the existing overburden pressure and the average increase of pressure) and the given soil parameters.

Input screen 1
The data elements for the embankment and surcharge properties, vertical drain design and the time frames of the project are defined in the input screen shown in Worksheet Display Screen 8.1a and 8.1c.

	A	B	C	D	E	F	G
1	EMBANKMANT ANALYSIS AND GROUND IMPROVEMENT TECHNIQUES						
2	------------	-------------	-------	-------	-------		
3		JOB DESCRIPTION :		**Coomera**			
4							
5			EMBANKMENT AND SURCHARGE PROPERTIES				
6	+++++++++++	+++++++++++	++++++++	+++++++	+++++++	++++++++	+ ++++
7	Design Height of Embankment (H)			2.5	metres		+
8	Embankment Width at top			5	metres		+ Full v
9	Slope of Embankment is 1 Vertical :			2	Horizontal		+ ==>
10	Analysis Increment surcharge Height			1	metres		+ {alph
11	Unit weight of placed fill			18	kN / cu metre		+ {alph
12							+
13			VERTICAL DRAIN DESIGN				+
14	+++++++++++	+++++++++++	++++++++	+++++++	+++++++	++++++++	+ ++++
15	(T)riangular or (S)quare pattern			s			+
16	(W)ick or (S)and Drains			s			+
17	Spacing of Drains S			3	metres		+ (Use
18	Radius of sand drain ,rd			0.15	metres		+ Equi
19	Smear Factor			1.5	* de		+ Equi
20							+ Factc

Worksheet 8.2a. Display screen.

	G	H	I	J	K	L	M
1							
2							
3							
4							
5	ERTIES		NOTES				
6	+	++++++++	++++++++	++++++++	++++++++	++++++++	+
7	+						+
8	+	Full width = 2 * B1					+
9	+	==> Slope width (B2) =			5	metres	+
10	+	{alpha1} =	0.0266	radians			+
11	+	{alpha2} =	1.5308	radians			+
12	+						+
13	+						+
14	+	++++++++	++++++++	++++++++	++++++++	++++++++	+
15	+						+
16	+						+
17	+	(Use S = 100 for no vertical drains)					+
18	+	Equivalent Spacing de			1.692	metres	+
19	+	Equivalent Drain Radius rw			0.15	metres	+
20	+	Factor de for smear effect			2.538	metres	+
21	+	n =	8.46		m =	1.419100	+

Worksheet 8.2b. Display screen.

The selection of sand or wick drains determines the calculation of the equivalent radius and screen displays.

At cell B18 the displays may vary between 'radius of sand drain, rd' or 'width of wick drain, b' while at cell B19 the display may be 'thickness of drain, t' or ' '.

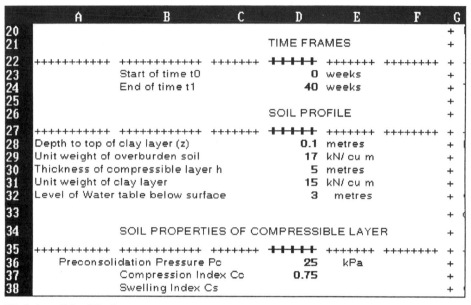

Worksheet 8.2c. Display screen.

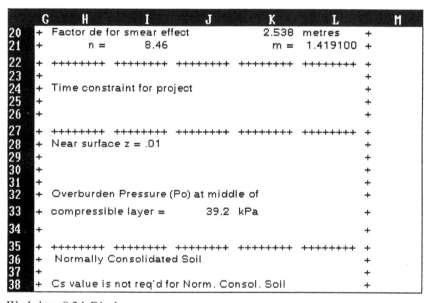

Worksheet 8.2d. Display screen.

	A	B	C	D	E	F	G
39	Initial Void Ratio eo			1.6			+
40	Coeff. of Consol.(Vertical) Cv			5	sq m / yr		+
41	Coeff. of Consol. (Radial) Cvr			10	sq m / yr		+
42	Undrained Cohesion			50	kPa		+
43	Bearing Capacity safety factor			2.5			+
44							+
45			DESIGN CONSIDERATIONS				+
46	+++++++++++ ++++++++++++ +++++++ +++++++ +++++++ ++++++++						+
47	The following should be noted :						+
48	– A Maximum Embankment Height (H) of			5.713	metres is required		+
49	to avoid Bearing capacity failure						+
50	– Embankment may be constructed to full Final Height						+
51	– A constant Side slope and embankment width assumed, and is						+
52	independent of surcharge height						+
53	– The Slope stability of the embankment should be checked						+
54	– Fill Height for settlemment analysis will be constrained by						+
55	* The maximum height which will not cause Bearing failure						+
56	* Side slopes (1V : H) and embankment width (2* B1) specified						+
57	– Influence factor is calculated to middle of the compressible						+
58	layer & middle of the embankment in the analysis below						+
59	– Average increase in pressure over the layer is used						+

Worksheet 8.2e. Display screen.

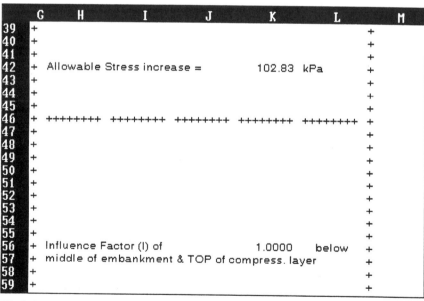

	G	H	I	J	K	L	M
39	+					+	
40	+					+	
41	+					+	
42	+	Allowable Stress increase =			102.83	kPa +	
43	+					+	
44	+					+	
45	+					+	
46	+	++++++++ ++++++++ ++++++++ ++++++++ ++++++++				+	
47	+					+	
48	+					+	
49	+					+	
50	+					+	
51	+					+	
52	+					+	
53	+					+	
54	+					+	
55	+					+	
56	+	Influence Factor (I) of			1.0000	below +	
57	+	middle of embankment & TOP of compress. layer				+	
58	+					+	
59	+					+	

Worksheet 8.2f. Display screen.

	A	B	C	D	E	F	G
61							+
62	*********	************	********	*******	*******	********	*
63					Time (weeks)		:
64					Time (years)		:
65					Radial Time factor Tr		:
66				Deg of radial consolidation Ur (%)			:
67					Vertical Time factor Tv		:
68				Deg of vertical consolidation Uv (%)			:
69			Degree of Consolidation (radial + vertical) %				:
70	*********	************	********	*******	*******	********	*
71							
72					CALCULATION OF INFLUENCE FAC
73	*********	************	*******	*******	*******	********	*
74		Height of	Slope	{ alpha 1 }			
75		Fill (metres)	Width	Middle	Top	Bottom	
76		h	(B2)				
77	*********	************	*******	*******	*******	********	*
78			2.5	5.00	0.4713	0.0266	0.5178
79			3.5	7.00	0.5379	0.0295	0.6223
80			4.5	9.00	0.5827	0.0313	0.6976
81			5.5	11.00	0.6147	0.0326	0.7538
82			6.5	13.00	0.6388	0.0335	0.7971
83			7.5	15.00	0.6575	0.0343	0.8314

Worksheet 8.2g. Display screen.

	G	H	I	J	K	L	M
61	+	CALCULATION OF CONSOLIDATION PARAMATERS				+	
62	*	********	********	********	********	********	*
63	:	0	10	20	30	40	*
64	:	0.00	0.19	0.38	0.58	0.77	*
65	:	0.00	0.30	0.60	0.90	1.19	*
66	:	0.0	81.4	96.5	99.4	99.9	*
67	:	0.00	0.15	0.31	0.46	0.62	*
68	:	0.0	44.3	62.1	74.0	82.2	*
69	:	0.0	89.6	98.7	99.8	100.0	*
70	*	********	********	********	********	********	*
71							*
72	ACTORS WITH DEPTH						*
73	*	********	********	********	********	********	*
74			INFLUENCE FACTOR (I)			Avg (kPa)	*
75			Middle	Top	Bottom	Pressure	*
76		{alpha2}=	0.7658	1.5308	0.4558	Increase	*
77	*	********	********	********	********	********	*
78			0.938	0.269	1.081	38.3	*
79			0.952	0.757	1.226	60.8	*
80			0.961	0.757	1.226	78.7	*
81			0.968	0.757	1.226	96.6	*
82			0.972	0.671	0.882	106.1	*
83			0.976	0.516	0.836	118.2	*

Worksheet 8.2h. Display screen.

	C	D	E	F	G	H	I	J	K	L
84	****	******	******	*********	*	******	******	******	******	*******	*
85										
86										
87										
88										
89							CALCULATION OF SETTLEMENTS with TIME			
90	*	******	******	*********	*	******	******	******	******	*******	*
91	*				│		T I M E (weeks)				*
92	*	Height	Average	Total	│	0	10	20	30	40	*
93	*	of	Pressure	Induced	│	------	------	------	------	-------	*
94	*	Fill	Increase	Settlement	│						*
95	*				│		Settlements (mm) expectecwith time			*
96	*	(metres)	(kPa)	(mm)	│		and v e r t i c a l d r a i n s			*
97	*	******	******	*********	*	******	******	******	******	*******	*
98	*	2.5	38.3	426.6	0	382	421	426	426	*
99	*	3.5	60.8	586.7	0	526	579	586	587	*
100	*	4.5	78.7	689.7	0	618	681	689	690	*
101	*	5.5	96.6	778.2	0	698	768	777	778	*
102	*	6.5	106.1	820.8	0	736	810	819	821	*
103	*	7.5	118.2	870.9	0	781	859	869	871	*
104	*	******	******	*********	*	******	******	******	******	*******	*

Worksheet 8.2i. Display screen.

	C	D	E	F	G	H	I	J	K	L
104	*	****	****	*******	*	*****	********	************	****	******	*
105										
106										
107						GRAPH AIDS to illustrate results				
108			\ \\\\\\\\\\\\\\ \\\ \\\\\\\\\ \\\\\\\\\\\\\\ \\\\\\\\\\\\\\\\\\\\\ \\\\\\\\\ \\\\\\\\\\ \							
109			\ Design Settlement line			for fill height of		2.5 metres	\
110			\							
111			\ " Dummy "	846		846		846	846	846 \
112			\ Lines	821		821		821	821	821 \
113			\							\
114			\		Drain Radius (mm)		150			\
115			\		Drain Spacing, m		3			\
116			\		Square pattern					\
117			\							\
118			\ \\\\\\\\\\\\\\ \\\ \\\\\\\\\ \\\\\\\\\\\\\\ \\\\\\\\\\\\\\\\\\\\\ \\\\\\\\\ \\\\\\\\\\ \							
119										

Worksheet 8.2j. Display screen.

This requires the input of the thickness of the wick drain or alternatively a blank display ' ' will occur.

Subsequently the equivalent drain radius r_w is calculated at cell K19.

Input screen 2

The data elements for the soil profile and the soil properties of the compressible layer are given in this input screen.

The calculation of the overburden pressure involves three possibilities for the water level:

1. Water below the middle of the clay layer → unit weights as input.

Cell (Format)	Display	Cell Entry Formula
B18	Radius of sand drain, rd	@IF(D16="w","Width of wick drain, b","Radius of sand drain,rd")
B19		@IF(D16="w", "Thickness of drain, t"," ")
K19		@IF(D16="w",(D18+D19)/2,D18)

Worksheet 8.2. Definition of drain type.

2. Water level below the overlying soil but above the middle of the clay layer → unit weight of overlying soil as input; effective unit weight calculated for part below the water level.

3. Water level above the overlying soil and the middle of the clay layer → effective unit weight of overlying soil for part below the water level; effective unit weight of clay.

At cells H32 and H33.

H32: 'Overburden pressure (P_o) at middle of
H33: 'compressible layer =
with the calculation at cell J33

The value of the overburden pressure is then compared to the preconsolidation pressure to determine the soil type (Table 8.1), and this is displayed in cell H36.

The swell index value is not required for normally consolidated soils and a note to this effect is provided in cell H38.

Cell (Format)	Display	Cell Entry Formula
J33	103.33	@IF(D28<D32, @IF(D28+D30/2<D32,+D28*D29+(D31*D30/2),D28*D29+D31*(D32−D28)+(D31−9.81)*(D28+D30/2−D32)),(D29*D32)+(D29−9.81)*(D28−D32)+(D31−9.81)*D30/2)
H36	Normally Consolidated soil	@IF(D36<=J33, " Normally Consolidated Soil","Over consolidated soil")
H38	Cs value is not req'd for Norm. Consol. Soil"	@IF(D36<=J33, "Cs value is not req'd for Norm. Consol. Soil","Cs is usually 5 to 10% of Cc value")

Worksheet 8.2. Definition of soil type.

No new techniques occur in the rest of the worksheet, except to integrate the various components of the analysis process. The reader should be able to follow the flow of calculations with the aid of the appendix listings when required.

The use of windows would be very appropriate in a large worksheet as this. The commands would be:

```
/ Worksheet Window Horizontal
```

The worksheet is then broken horizontally at the cursor position.

Movement between windows is accomplished by the {F6} function key.

The graphical output is shown in Graphsheet 8.2 and is generated using the graph aids as explained with Spreadsheet 8.1.

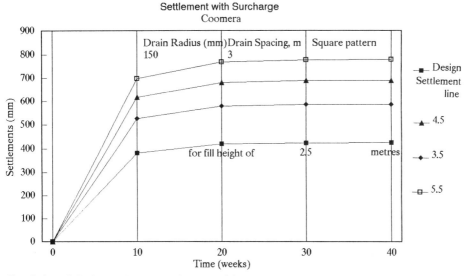

Graphsheet 8.2. Consolidation settlement with surcharge and vertical drains.

In Lotus 3.X, the current graph can be displayed simultaneously with the worksheet using:

```
/ Worksheet Window Graph
```
This is useful in 'what-if' graphing.

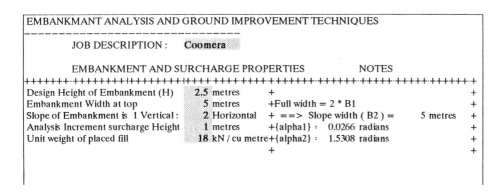

```
                    VERTICAL DRAIN DESIGN   +                                    +
+++++++·+++++++++++H+++++++++++H+++·++++·++++·+++++++·++++++·+++++·+++++·++++++·+
(T)riangular or (S)quare pattern        s           +                            +
       (W)ick or (S)and Drains          s           +                            +
       Spacing of Drains S            3 metres      +( Use S = 100 for no vertical drains )     +
       Radius of sand drain,rd     0.15 metres      +Equivalent Spacing de        1.692 metres  +
       Smear Factor                 1.5 * de        +Equivalent Drain Radius rw    0.15 metres   +
                                                    +Factor de for smear effect   2.538 metres   +
                    TIME FRAMES                     +     n =     8.46        m =    1.4191 +
+++++++·+++++++++++H+++++++++++H+++·++++·++++·+++++++·++++++·+++++·+++++·++++++·+
       Start of time t0               0 weeks       +                            +
       End of time t1                40 weeks       +Time constraint for project +
                                                    +                            +
                    SOIL PROFILE                    +                            +
+++++++·+++++++++++H+++++++++++H+++·++++·++++·+++++++·++++++·+++++·+++++·++++++·+
Depth to top of clay layer (z)      0.1 metres      +Near surface z = .01                        +
Unit weight of overburden soil       17 kN/ cu m    +                            +
Thickness of compressible layer h     5 metres      +                            +
Unit weight of clay layer            15 kN/ cu m    +                            +
Level of Water table below surface    3 metres      +Overburden Pressure (Po) at middle of       +
                                                    +compressible layer =   39.2 kPa             +
SOIL PROPERTIES OF COMPRESSIBLE LAYER               +                            +
+++++++·+++++++++++H+++++++++++H+++·++++·++++·+++++++·++++++·+++++·+++++·++++++·+
       Preconsolidation Pressure Pc    25 kPa       + Normally Consolidated Soil                 +
              Compression Index Cc    0.75          +                            +
              Swelling Index Cs                     +Cs value is not req'd for Norm. Consol. Soil +
Initial Void Ratio eo                 1.6          +                            +
Coeff. of Consol.( Vertical ) Cv        5 sq m / yr +                            +
Coeff. of Consol. ( Radial ) Cvr       10 sq m / yr +                            +
Undrained Cohesion                     50 kPa       +Allowable Stress increase =   102.83 kPa     +
Bearing Capacity safety factor        2.5          +                            +
                                                    +                            +
```

Worksheet form 8.1. Data entry and output table. Filename: Embank1.

```
CONSOLIDATION WITH TIME, FILL HEIGHT & VERTICAL DRAINS
----------------------------------------------------------------
                JOB TITLE :              Coomera
              SOIL AND SURCHARGE PROPERTIES
+++++++++++++++++++++++++++++++++++++++++++++++++++
Coefficient of compressibility Mv              0.001 sq m /kN      +
Coeff. of Consolidation (Vertical) Cv              5 sq m /yr      +
Coeff. of Consolidation (Radial) Cvr              10 sq m /yr      +
Thickness of compressible clay layer H             5 metres       +
Unit weight of placed fill                        18 kN/cu m      +
Design Height of Fill                            2.5 metres       +
Increment Fill Height                              1 metres       +
              PROPERTIES OF VERTICAL DRAINS                       +
+++++++++++++++++++++++++++++++++++++++++++++++++++
              Drain Radius rd                   0.15 metres       +
              Spacing of Drains S                  3 metres       +
(S)quare or (T)riangular pattern          s                       +
              ==> Equivalent radius (R) =                  1.692  +
Factor to account for smear :            1.5 * R            2.538  +
                      =====>          n =                   8.46   +
                      =====>          m =             1.41910009   +
```

TIME FRAMES					+
Start of time t0		0 weeks		+	
End of time t1		40 weeks		+	
Time (weeks)	0	10	20	30	40 +
Time (years)	0.00	0.19	0.38	0.58	0.77 +
Radial Time factor Tr	0.00	0.30	0.60	0.90	1.19 +
Deg of radial consol Ur (%)	0.0	81.4	96.5	99.4	99.9 +
Vertical Time factor Tv	0.00	0.15	0.31	0.46	0.62 +
Deg of vert consol Uv (%)	0.0	44.3	62.1	74.0	82.2 +
Deg of Consol (rad+vert)%	0.0	89.6	98.7	99.8	100.0 +

Height of Fill (metres)	Total	Settlements (mm) expected with vertical drains			+ +
2.5	225.0	0.0	201.7	222.1	224.6	225.0 +
3.5	315.0	0.0	282.4	310.9	314.5	314.9 +
4.5	405.0	0.0	363.1	399.7	404.3	404.9 +
5.5	495.0	0.0	443.7	488.5	494.2	494.9 +
6.5	585.0	0.0	524.4	577.3	584.0	584.9 +
7.5	675.0	0.0	605.1	666.2	673.9	674.9 +

Graph Aids to help illustrate the results
Design Settlement line   for fill height of   2.5 metres

Text illustrating line	675	675	675	675	675
	650	650	650	650	650

Drain Radius (mm)	150
Drain Spacing (m)	3
Square pattern	

Worksheet form 8.2. Data entry and output table. Filename: Embank2.

```
++++++++++++++++++++++++++++++++++++++++++++++++++++++++++++++++++++++++
              DESIGN CONSIDERATIONS                                    +
++++++++++++++++++++++++++++++++++++++++++++++++++++++++++++++++++++++++
```

The following should be noted :
- A Maximum Embankment Height (H) of   5.713 metres is required
  to avoid Bearing capacity failure
- Embankment may be constructed to full Final Height
- A constant Side slope and embankment width assumed, and is
  independent of surcharge height
- The Slope stability of the embankment should be checked
- Fill Height for settlement analysis will be constrained by
  * The maximum height which will not cause Bearing failure
  * Side slopes (1V : H) and embankment width (2* B1) specified
- Influence factor is calculated to middle of the compressible
  layer & middle of the embankment in the analysis below
- Average increase in pressure over the layer is used

Influence Factor (I) of   1.0000   below
middle of embankment & TOP of compress. layer

CALC. OF CONSOLIDATION PARAMATERS

		0	10	20	30	40
Time (weeks)	:	0	10	20	30	40 *
Time (years)	:	0.00	0.19	0.38	0.58	0.77 *
Radial Time factor Tr	:	0.00	0.30	0.60	0.90	1.19 *
Deg of radial consolidation Ur (%)	:	0.0	81.4	96.5	99.4	99.9 *
Vertical Time factor Tv	:	0.00	0.15	0.31	0.46	0.62 *
Deg of vertical consolidation Uv (%)	:	0.0	44.3	62.1	74.0	82.2 *
Degree of Consolidation (radial + vertical) %	:	0.0	89.6	98.7	99.8	100.0 *

CALCULATION OF INFLUENCE FACTORS WITH DEPTH

Height of Fill (metres) h	Slope Width (B2)	{ alpha1 } Middle	Top	Bottom	INFLUENCE FACTOR (I) Middle	Top	Bottom	Avg (kPa) Pressure Increase
					{alpha2} = 0.7658	1.5308	0.4558	
2.5	5.00	0.4713	0.0266	0.5178	0.938	0.269	1.081	38.3 *
3.5	7.00	0.5379	0.0295	0.6223	0.952	0.757	1.226	60.8 *
4.5	9.00	0.5827	0.0313	0.6976	0.961	0.757	1.226	78.7 *
5.5	11.00	0.6147	0.0326	0.7538	0.968	0.757	1.226	96.6 *
6.5	13.00	0.6388	0.0335	0.7971	0.972	0.671	0.882	106.1 *
7.5	15.00	0.6575	0.0343	0.8314	0.976	0.516	0.836	118.2 *

Worksheet form 8.2 (continued).

CALCULATION OF SETTLEMENTS with TIME

Height of Fill	Average Pressure Increase	Total Induced Settlement		TIME (weeks)				
				0	10	20	30	40
				Settlements (mm) expected and vertical drains				with time
(metres)	(kPa)	(mm)						
2.5	38.3	426.6		0	382	421	426	426
3.5	60.8	586.7		0	526	579	586	587
4.5	78.7	689.7		0	618	681	689	690
5.5	96.6	778.2		0	698	768	777	778
6.5	106.1	820.8		0	736	810	819	821
7.5	118.2	870.9		0	781	859	869	871

GRAPH AIDS to illustrate results

Design Settlement line		for fill height of		2.5 metres	
" Dummy "	846	846	846	846	846
Lines	821	821	821	821	821
	Drain Radius (mm)		150		
	Drain Spacing, m		3		
	Square pattern				

Worksheet form 8.2 (continued).

CHAPTER 9

Macro programming

Macros were originally described as 'typing alternatives'. In its most basic form a macro is simply a collection of keystrokes which have been 'programmed' into a two key sequence. Macros therefore save time and increase efficiency by automating the more frequently used keystrokes.

The worksheets developed in earlier chapters may also have been produced as a macro program with iterative call routines instead of replication of the basic formula into various cell ranges.

The worksheets may also be menu driven in order to provide a more user friendly interface.

The worksheets of Chapter 3 will be redeveloped here to illustrate the fundamentals of macro programming.

9.1 APPLICATION OF MACROS

Macros are entered into cells as a label, whether the macro is a command, number or formula. Lotus 123 reads the macro from left to right until it reaches a vacant cell or hard return (~) then continues below the left most cell of the series of macro commands.

Macros may be created in any part of the worksheet, but is usually placed so as not to affect the calculating worksheet area, such as in the AA column. The macros may be documented in adjacent cells for future reference or explanations. This will not affect the program flow, but can be considered as a REMark type statement.

Macros are named by issuing the command:

`/ Range Name Create` {Macro Name}
{Macro Range}

The macro name must be preceded by a backslash \ and any letter of the alphabet except O. The macro range is the cell entry at which the macro starts execution. The macro is then executed by using the ALT + {MACRO NAME}.

EXAMPLE: ALT + P keys pressed together may be used to represent the printing sequence for a worksheet.

In Chapter 3 the worksheet occupied cells A1.F20, and an analysis may involve evaluation of stresses for varying distances (*x* or *y*) from the load. A printout may be required for each condition. This would involve using the following command for each hardcopy: / Print Printer Range {A1.F20} Go.

To set up the macro the procedure is:
- At cell AA1 enter: ' / PPRA1.F20G,
- Name the macro: \P range AA1.

To execute the macro: Press ALT + P simultaneously.

The worksheet is then printed.

Macros usually involve extensive cell referencing, either to examine a cell entry or to place a cell entry. This is accomplished by using the following commands for cursor movement:

{Goto} {Left} {Right 4} {Up} {Down 2}, etc.

The reader is referred to the Lotus 123 manual for further explanations and examples.

Some examples of the uses of macros are given here with respect to the spreadsheet of Chapter 3. Some common macro applications are:
- Table generating,
- Remote calculations,
- Menus.

Table generating
Worksheet 3.1 analysed stresses due to a point load. The worksheet was developed using the Copy commands in order to replicate the formula into adjacent cells. Thus the tabled results are produced with formulae in each cell converted to a numerical result.

A tabular result may also be produced by using the formula *once*, but having an iterative access until the range of results are computed. In this way the output of the cells would be a calculated numeric value, without the need to have the formula repeated in each cell entry. This may be accomplished by two ways:
1. Use of the / Data Table command,
2. Use of macros.

Remote calculations
Instead of the formula being produced in the main worksheet screen, the calculations may be performed 'remote' to that screen and the calculated values copied to the main screen. This practically eliminates the incidence of the user erroneously tampering with any of the formulae. (The use of the Protect feature provides only limited protection to the exploratory user).

Menus

Menus may be developed and displayed in the control panel and then used in a similar way as the 1-2-3 selection menu.

Up to eight individual programs may be initiated from this menu-selection structure. This acts as a user friendly environment to the user, and may be combined with any of the above mentioned features (remote calculations).

These will now be applied in subsequent sections.

9.2 MACRO FOR TABLE GENERATING AND REMOTE CALCULATIONS

A macro is used here for calculating the stress induced from a point load (Chapter 3). All calculations are performed remote from the main worksheet screen and the values (not formula) are copied to the main screen.

Worksheet 9.1 shows the basic data entry required. The macro is developed separately in the AA and BA columns and is output in a similar fashion as that shown in Chapter 3.

Worksheet 9.1 shows the screen prompts which occurs in a window of the worksheet. To create the window, place the cursor at the desired position (row 10) and use:

`/ Worksheet Window Horizontal`

Movement from window to window is achieved by using the {F6} function key. The screen prompt message is then typed into the relevant cells. Depending on the

	A	B	C	D	E	F
1			STRESS VARIATION DUE TO POINT LOAD			
2						
3			Surface Point Load (Q) = ?		100	<- kN
4			Analysis begins at a depth = ?		10	<- metres
5			Final Depth (z) = ?		20	<- metres
6			No of increments (N) = ?		1	<-
7		x - Distance from centre of loading = ?			1	<- metres
8		y - Distance from centre of loading = ?			1	<- metres
9						
10						
11						
30	A	B	C	D	E	F
31	The highlighted cells are the data entry cells. Enter data					
32	in the highlighted cells (Cells E3 to E8). The worksheet					
33	calculates the Vertical & Shear stress variation with depth.					
34						
35	FILL in the above data and press ALT + A to activate macro					
36	The macro will take a few seconds to execute, during which					
37	CMD will appear at the bottom centre of the screen.					
38						

Worksheet 9.1. Data entry input form with window.

position of the cursor when saved, the message (in the window) appears each time the worksheet is retrieved and remains there as an aid to the user.

Since this manual is intended to deemphasise programming in preference to the use of 'quick and easy' customised programs, only a cursory treatment of the applications is shown and detailed explanations of the macro worksheet development is not given. The reader is directed to the Lotus 123 Reference Manual or one of the specialised books on macro programming.

The macro is named as follows:

```
/ Range Name Create \A AA1
```

The macro is named 'A' and begins execution at cell AA1.

The listing of the macro in the worksheet now follows:

The main macro (\A) has a few named macros as branches. These form part of the iteration sequence of commands in the macro.

```
AA1: '{paneloff}{windowsoff}
```

This suppresses the continuous screen scrolling during the execution of the macro. The control panel and the screen display are suppressed. After invoking the macro the user is aware of the macro execution by the CMD displayed at the bottom of the screen.

Note that cursor commands are bracketed { } in the macro.

The named macros are:

```
AA2: '/rncloop~ag21~
AA3: '/rncend~ag24~
AA4: '/rnccont~aa21~
AA5: '/rnccarryon~aa23~
```

This calls the Lotus menu (/) and uses the commands:

```
/ Range Name Create {Name of macro} {Location of macro}
```

A word is used for naming the macros in this instance (loop, end, cont, carryon).

The rest of the macro covers the screen formatting, positioning for execution, table headings and the formulae required.

```
AA6: '{goto}be9~
AA7: '@sqrt(e7^2+e8^2)~/rff3~~ /rv~e9~
AA8: '{goto}f9~metres~{left 4}' =====> ~
AA9: '{right}'r = Distance = ~
```

This part of the macro calculates the *r*-distance at cell BE9 then copies this calculated *value* to the main worksheet area at cell E9.

```
AA10: '{down}{left 2}\+~/c~{right}.{right 4}~
AA11: '{down}^Depth~{right}^Least Distance~
AA12: '{right}^VERTICAL~{right}^ SHEAR~
```

AA13: '{down}^STRESS{{}kPa{}}~{left}^ STRESS{{}kPa{}}~
AA14: '{left}^R{{}m{}}~{left}^metres~
AA15: '{down}\+~/c~{right}.{right 4}~

The table heading is set up here.

AA16: '{goto}ba14~+e4~/rv~a14~
AA17: '{right}@sqrt(e9^2+a14^2)~/rv~b14~
AA18: '{right}{3*e3*(a14)^3)/(2*@pi*(b14)5)~/rv~c14~
AA19: '{right}{3*e3*(a14)^2*e9)/(2*@pi*(b14)^5)~/rv~d14~

The initial depth is used to calculate the least distance (R), and the vertical and shear stresses. This is calculated at a remote location (cells BA14 to BE14). The calculated *values* are copied to the main worksheet area.

AA20: '{goto}ba14~
AA21: '{if@cellpointer("contents")>=+e5}{branch end}
AG21: '{down}+{up}+((e5-e4)/e6)~

The cell pointer is moved to cell BA14 and the contents of this cell are checked with the final depth (E5). If this is the final depth the table stops here. If it is not the final depth, the cell pointer moves down and increments the depth.

AA22: '{branch loop}
AG22: '{branch cont}
AA23: '{end}{up}/rff3~.{end}{down}~/rv.{end}{down}~a14~
AA24: '{end}{down 2}{right}99~/c~{right}.{right}~
AG24: '{branch carryon}

The process is iterated and the tabled depths produced until the final depth is reached. The calculated values are formatted to 3 decimal places.

AA25: '{end}{up}/c~{down}.{end}{down}{up}~{end}{down}/
    re~{end}{up}{end}{up}/rff3~.{end}{down}~/rv.{end}
    {down}~b14~
AA26: '{right}/c~{down}.{end}{down}{up}~{end}{down}/re~{end}
    {up}{end}{up}/rff2~.{end}{down}~/rv.{end}{down}~c14~
AA27: '{right}/c~{down}.{end}{down}{up}~{end}{down}/re~{end}
    {up}{end}{up}/rff2~.{end}{down}~/rv.{end}{down}~d14~

This copies the formulae to the full table range of depths. The values are formatted and copied to the main worksheet.

AA28: '{Home}{down 20}{beep}

This signals the end of the macro execution with a beep sound. The screen moves to the tabled results. The CMD in the control panel disappears.

The output table is the same as in Chapter 3. The graph may also be produced if required.

A program written with the use of macros would usually be designed to perform calculations in the background. This method requires less memory and therefore a more efficient program is produced.

However, it readily becomes evident that writing of these macros requires programming skill, unlike the non macro spreadsheet developments where the formula is replicated.

The philosophy in this manual has been to provide a tool that is 'quick and easy' to use and which does not require specialist time. Therefore the non macro spreadsheet approach is recommended. There are times when programming is necessary (specialised applications) and macros provide that facility.

9.3 CUSTOMISED MENUS

Simple customised menus may be developed to aid in repetitive features such as printing specific field ranges, choosing between spreadsheets, etc. Figure 9.1 shows the menu structure of Worksheet 9.2 and the listing is provided in the appendix.

The actual displays are shown on Worksheet Screens 9.2 and 9.3.

The macro begins at cells GI2312:

GI2312: `'\0`
GI2313: `'\m`

This names the macro.

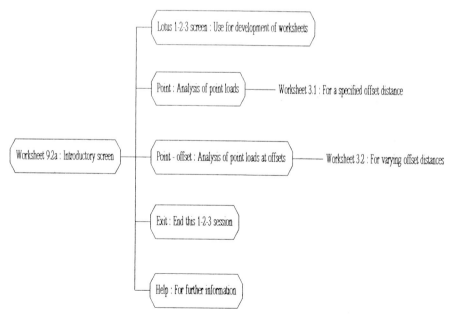

Figure 9.1. Menu structure.

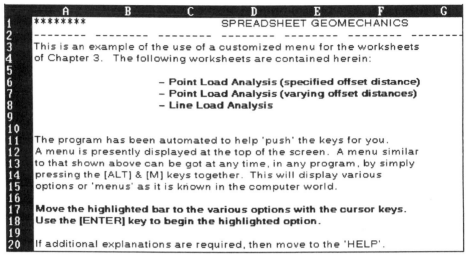

Worksheet 9.2a. Display screen.

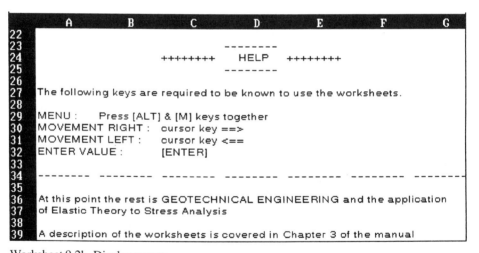

Worksheet 9.2b. Display screen.

Worksheet 9.2 is retrieved with the screen message shown as well as a menu command. The \0 range name is used to automatically call the menu on retrieving the spreadsheet. The menu may be accessed at any time by pressing the ALT + M keys. This executes the \m range name. Both macro ranges precede the menu-branch which controls the macro execution.

GJ2313: '{menubranch gh2316}~

On retrieving the spreadsheet Worksheet Screen 9.2a is displayed with the menu. The menu is entered as follows:

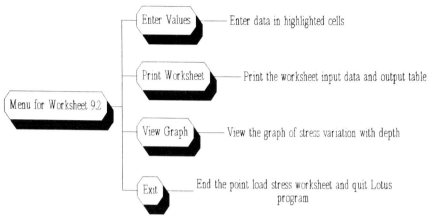

Figure 9.2. Menu display.

GH2316: `'LOTUS 123`
GI2316: `^POINT`
GJ2316: `^POINT-OFFSET`
GK2316: `"LINE` (optional for a line load worksheet)
GL2316: `'EXIT`
GM2316: `'HELP`

Just as in the Lotus 123 development menu, the options are accessed by moving the cursor (or typing the first letter) then the various sub-menus would be displayed.

Below each of the options the sub-menus would be displayed by entering in the cells directly below the relevant option:

GH2317: `'Use the worksheet for development or for applications not shown in this menu`
GI2317: `'Analysis of point load stresses for a specified offset distance`
GJ2317: `'Analysis of point load stresses for varying offset distances`
GK2317: `'Analysis of line load stresses`
GL2317: `'End this program and return to main menu – "Have a nice day"`
GM2317: `'For additional assistance other than given herein contact Burt Look`

Up to this point the above shows only the display menu at the top of the screen. The actual program (keys to be pushed) are entered below each menu / sub menu option.

> EXAMPLE: The cursor is positioned at the Lotus 123 option. This erases the screen display for development work. The commands are:
>
> Menu Display at GH2316: `Lotus 123`
> Sub Menu Display at GH2317: `Use the worksheet for development for applications not shown in this menu`
> Execution commands at GH2318: `/WEY~`
>
> Without the macro this would involve: `/Worksheet Erase Yes`

GH2318: `'/WEY~`
GI2318: `'/FRmacstrp~`
GJ2318: `'/FRmacstrq~`
GK2318: `'/FRmacstrl~`
GL2318: `'/QYE~`
GM2318: `'{goto}a22~`
GM2319: `'{menucall gh2316}~`

Another utility option is the HELP menu. HELP would display the worksheet of 9.2b.

If the POINT worksheet was required and this option selected then the commands to be executed would be:

`/FRmacstrp~`

This retrieves the File Macstrp. This is file 'Stressp' of Chapter 3 with its own customised menu.

Worksheet 9.3 is displayed with a window. This is a simple utility menu to automate some of the more used functions.

The full macro listing to produce the above screen displays is given in the appendix. The macro is entered at cells GI2316 to GL2318.

	A	B	C	D	E	F
1	STRESS VARIATION DUE TO POINT LOAD					
2	--------	-------------	-------------	--------		
3						
4		Surface Point Load (Q) = ?			100	kN
5		Analysis begins at a depth = ?			1	metres
6		Increment depth (z) = ?			0.5	metres
7	x – Distance from centre of loading = ?				1	metres
8	y – Distance from centre of loading = ?				1	metres
29	A	B	C	D	E	F
30	PRESS {ALT + M} KEYS TOGETHER TO VIEW MENU					
31						
32	The highlighted cells seen above are the data entry cells.					
33	You can enter values only in these highlighted cells. The					
34	worksheet then calculates the stress variation with depth.					
35	The vertical and shear stress distributions are calculated					
36	for the specified offset distance					
37						

Worksheet 9.3. Display screen with window.

It is evident that macros would fall into two main categories:
1. To automate spreadsheets,
2. As a programming language.

Programs which are commercially available use the macro facilities to automate the analysis procedure e.g. PISSAP (Randolph 1989) is a Lotus 123 spreadsheet which automates simple elastic design calculations for single piles. This program provides graphical output with cartoon displays and is menu driven.

The recent Lotus release uses Smarticons to perform various commands. It can also be used to run macros. There are twelve user icons available to assign a macro and a description of the icon.

Activate the Smarticon with clicking on with a mouse or alternatively use the keyboard: ALT + F7.

Go to palette 6 and click on to the icon which assigns a macro.

The macros can then be assigned to one of the twelve user defined icons.

CHAPTER 10

Spreadsheet as a design tool

The engineer has at his disposal some very simple and easy to use design 'WHAT IF' tools, which may be used to examine cost effective designs. Experience will still be essential to check the practicality of the illustration and to check that the numbers make sense in the first instance. A number is NOT a solution unless applied appropriately.

This is essentially a powerful management tool to examine alternative strategies. The total geotechnical design is for the avoidance of the single number trap with all its limitations and to be able to provide a range of conditions.

Alternative analysis and design approaches have traditionally been limited because of the time consuming mechanics of the design process. Spreadsheets can now break these limitations. An essentially simple deterministic model is produced while providing a pseudo effect of a stochastic design. What-if experimenting becomes a simple exercise when formula arguments (data) are placed in separate cells and those cells are referenced, by the formula for a range of conditions (tabled output).

The more recent releases of Lotus 2.X, 3.X and 4.X for Windows provide added facilities to the spreadsheet design. Useful facilities for engineering purposes include:

– Multiple worksheets,
– Improved graphing, with combined bar and line graphs, area charts, etc.,
– Ability to view graphs with the Lotus worksheet simultaneously. This provides an instantaneous graphical 'what-if' design approach,
– Add-in programs which extend the capabilities of 123. These include: AUDITOR which indentifies formula relationships and can help you to find worksheet formula errors; WYSIWYG which creates a presentation-quality worksheet print out including graphs and data on the same page; VIEWER which displays the file name and its contents in a split screen and allows you to browse through the contents and link this file to other 123 worksheet files.
– Third party add-in products for 123. A notable example worth exploring for geotechnical engineering applications is @Risk. This program performs risk analysis and simulation modelling. @Risk allows you to define uncertain cell

values in 123 as probability distributions using @ functions.

Both Monte Carlo and Latin Hypercube sampling techniques are supported. The `@Risk Graph` portion generates a graphic probability distribution of possible results for each output cell selected.

– Statistical functions, including regression analysis.

– Data analysis using data functions. Lotus Improv for Windows uses categories to group data. This enables the data to be examined from different perspectives – a useful tool for data analysis.

The spreadsheet has its limitations, such as in its application to the analysis of a multi-layered soil profile. In such an application, the spreadsheet must be designed to analyse a specific number of layers. Once set up in this manner, the user is constrained to the number of layers in the original design. The use of macros, as illustrated in Chapter 8, would alleviate this problem.

While this is a manual about solving geotechnical problems with programming tools, the philosophy has been to de-emphasise the programming aspects. The use of development tools are recommended rather than writing a program in a programming language.

The method of analysis is often secondary to the importance of selecting representative soil parameters and profiles. The model defined in the spreadsheet is used to identify the most sensitive aspects of the analysis. The ability to accomodate complex models in multi-layered systems is therefore not as important as the geotechnical idealisation which foreruns the analysis.

An approximate solution to the right problem is therefore more desirable than a precise solution to the wrong problem. The sensitivity analysis performed in the spreadsheet assists in defining the sensitive variables. Once isolated and the soil idealisation refined, the definition of the correct problem can at least be covered in the analysis and design.

Spreadsheets developed here are working models, but can be refined considerably and the screen presentation/user friendliness can be enhanced. It is hoped that these models may be used as a stepping stone to an even more refined and integrated analytical tool.

APPENDIX

Program listings

1. FILE: STRESSP

```
A1: PR 'STRESS VARIATION DUE TO POINT LOAD
A2: PR \-
B2: PR [W14] \-
C2: PR [W13] \-
B4: PR [W14] ' Surface Point Load ( Q ) = ?
E4: U 100
F4: PR 'kN
B5: PR [W14] 'Analysis begins at a depth = ?
E5: U 1
F5: PR 'metres
B6: PR [W14] '    Increment depth ( z ) = ?
E6: U 0.5
F6: PR 'metres
A7: PR 'x - Distance from centre of loading = ?
E7: U 1
F7: PR 'metres
A8: PR 'y - Distance from centre of loading = ?
E8: U 1
F8: PR 'metres
B9: PR [W14] "=====>
C9: PR [W13] ' r - Distance  =
E9: (F3) PR @SQRT(E7^2+E8^2)
F9: PR 'metres
A10: PR \+
B10: PR [W14] \+
C10: PR [W13] \+
D10: PR \+
E10: PR \+
A11: PR 'Depth
B11: PR [W14] 'Least Distance
C11: PR [W13] ^VERTICAL
D11: PR ^SHEAR
A12: PR '{m}
B12: PR [W14] ^R {m}
C12: PR [W13] 'STRESS {kPa}
D12: PR 'STRESS {kPa}
A13: PR \+
B13: PR [W14] \+
C13: PR [W13] \+
D13: PR \+
```

E13: PR \+
A14: (F2) PR +E5
B14: (F2) PR [W14] @SQRT(+A14^2+$E$9^2)
C14: (F2) PR [W13] 3*$E$4*A14^3/(2*@PI*B14^5)
D14: (F2) PR 3*$E$4*$E$9*A14^2/(2*@PI*B14^5)
A15: (F2) PR +A14+$E$6
B15: (F2) PR [W14] @SQRT(+A15^2+$E$9^2)
C15: (F2) PR [W13] 3*$E$4*A15^3/(2*@PI*B15^5)
D15: (F2) PR 3*$E$4*$E$9*A15^2/(2*@PI*B15^5)
A16: (F2) PR +A15+$E$6
B16: (F2) PR [W14] @SQRT(+A16^2+$E$9^2)
C16: (F2) PR [W13] 3*$E$4*A16^3/(2*@PI*B16^5)
D16: (F2) PR 3*$E$4*$E$9*A16^2/(2*@PI*B16^5)
A17: (F2) PR +A16+$E$6
B17: (F2) PR [W14] @SQRT(+A17^2+$E$9^2)
C17: (F2) PR [W13] 3*$E$4*A17^3/(2*@PI*B17^5)
D17: (F2) PR 3*$E$4*$E$9*A17^2/(2*@PI*B17^5)
A18: (F2) PR +A17+$E$6
B18: (F2) PR [W14] @SQRT(+A18^2+$E$9^2)
C18: (F2) PR [W13] 3*$E$4*A18^3/(2*@PI*B18^5)
D18: (F2) PR 3*$E$4*$E$9*A18^2/(2*@PI*B18^5)
A19: (F2) PR +A18+$E$6
B19: (F2) PR [W14] @SQRT(+A19^2+$E$9^2)
C19: (F2) PR [W13] 3*$E$4*A19^3/(2*@PI*B19^5)
D19: (F2) PR 3*$E$4*$E$9*A19^2/(2*@PI*B19^5)
A20: (F2) PR +A19+$E$6
B20: (F2) PR [W14] @SQRT(+A20^2+$E$9^2)
C20: (F2) PR [W13] 3*$E$4*A20^3/(2*@PI*B20^5)
D20: (F2) PR 3*$E$4*$E$9*A20^2/(2*@PI*B20^5)

2. FILE: STRESSQ

A1: PR [W10] 'STRESS VARIATION DUE TO POINT LOAD
A2: PR [W10] \-
B2: PR [W11] \-
C2: PR [W8] \-
D2: PR [W8] \-
A3: PR [W10] ' Surface Point Load (Q) = ?
E3: U [W7] 100
F3: PR [W7] 'kN
A4: PR [W10] 'Initial depth from surface = ?
E4: U [W7] 1
F4: PR [W7] 'metres
A5: PR [W10] '   Increment depth (z) = ?
E5: U [W7] 0.5
F5: PR [W7] 'metres
A6: PR [W10] 'x - Distance from centre of loading
E6: U [W7] 1
F6: PR [W7] 'metres
A7: PR [W10] 'y - Distance from centre of loading
E7: U [W7] 1
F7: PR [W7] 'metres
B8: PR [W11] ' Offset distance (r) = ?
E8: (F3) PR [W7] @SQRT(E6^2+E7^2)
F8: PR [W7] 'metres

B9: PR [W11] 'Offset increment (dr) = ?
E9: U [W7] 0.5
F9: PR [W7] 'metres
A10: PR [W10] \+
B10: PR [W11] \+
C10: PR [W8] \+
D10: PR [W8] \+
E10: PR [W7] \+
F10: PR [W7] \+
G10: PR \+
A11: PR [W10] 'Offset r->
B11: (F1) PR [W11] +E8-E9
C11: (F1) PR [W8] +E8
D11: (F1) PR [W8] +E8+E9
E11: (F1) PR [W7] +E8-E9
F11: (F1) PR [W7] +E8
G11: (F1) PR +E8+E9
A12: PR [W10] 'Depth z
B12: PR [W11] ' VERTICAL STRESS {kPa}
E12: PR [W7] '   SHEAR STRESS {kPa}
A13: PR [W10] \+
B13: PR [W11] \+
C13: PR [W8] \+
D13: PR [W8] \+
E13: PR [W7] \+
F13: PR [W7] \+
G13: PR \+
A14: (F1) PR [W10] +E4
B14: (F2) PR [W11] 3*$E$3*A14^3/(2*@PI*($B$11^2+A14^2)^(5/2))
C14: (F2) PR [W8] 3*$E$3*A14^3/(2*@PI*($C$11^2+A14^2)^(5/2))
D14: (F2) PR [W8] 3*$E$3*A14^3/(2*@PI*($D$11^2+A14^2)^(5/2))
E14: (F2) PR [W7] 3*$E$3/(2*@PI)*($E$11*A14^2/($E$11^2+A14^2)^(5/2))
F14: (F2) PR [W7] 3*$E$3/(2*@PI)*($F$11*A14^2/($F$11^2+A14^2)^(5/2))
G14: (F2) PR 3*$E$3/(2*@PI)*($G$11*A14^2/($G$11^2+A14^2)^(5/2))
A15: (F1) PR [W10] +A14+$E$5
B15: (F2) PR [W11] 3*$E$3*A15^3/(2*@PI*($B$11^2+A15^2)^(5/2))
C15: (F2) PR [W8] 3*$E$3*A15^3/(2*@PI*($C$11^2+A15^2)^(5/2))
D15: (F2) PR [W8] 3*$E$3*A15^3/(2*@PI*($D$11^2+A15^2)^(5/2))
E15: (F2) PR [W7] 3*$E$3/(2*@PI)*($E$11*A15^2/($E$11^2+A15^2)^(5/2))
F15: (F2) PR [W7] 3*$E$3/(2*@PI)*($F$11*A15^2/($F$11^2+A15^2)^(5/2))
G15: (F2) PR 3*$E$3/(2*@PI)*($G$11*A15^2/($G$11^2+A15^2)^(5/2))
A16: (F1) PR [W10] +A15+$E$5
B16: (F2) PR [W11] 3*$E$3*A16^3/(2*@PI*($B$11^2+A16^2)^(5/2))
C16: (F2) PR [W8] 3*$E$3*A16^3/(2*@PI*($C$11^2+A16^2)^(5/2))
D16: (F2) PR [W8] 3*$E$3*A16^3/(2*@PI*($D$11^2+A16^2)^(5/2))
E16: (F2) PR [W7] 3*$E$3/(2*@PI)*($E$11*A16^2/($E$11^2+A16^2)^(5/2))
F16: (F2) PR [W7] 3*$E$3/(2*@PI)*($F$11*A16^2/($F$11^2+A16^2)^(5/2))
G16: (F2) PR 3*$E$3/(2*@PI)*($G$11*A16^2/($G$11^2+A16^2)^(5/2))
A17: (F1) PR [W10] +A16+$E$5
B17: (F2) PR [W11] 3*$E$3*A17^3/(2*@PI*($B$11^2+A17^2)^(5/2))
C17: (F2) PR [W8] 3*$E$3*A17^3/(2*@PI*($C$11^2+A17^2)^(5/2))
D17: (F2) PR [W8] 3*$E$3*A17^3/(2*@PI*($D$11^2+A17^2)^(5/2))
E17: (F2) PR [W7] 3*$E$3/(2*@PI)*($E$11*A17^2/($E$11^2+A17^2)^(5/2))
F17: (F2) PR [W7] 3*$E$3/(2*@PI)*($F$11*A17^2/($F$11^2+A17^2)^(5/2))
G17: (F2) PR 3*$E$3/(2*@PI)*($G$11*A17^2/($G$11^2+A17^2)^(5/2))
A18: (F1) PR [W10] +A17+$E$5
B18: (F2) PR [W11] 3*$E$3*A18^3/(2*@PI*($B$11^2+A18^2)^(5/2))

C18: (F2) PR [W8] 3*$E$3*A18^3/(2*@PI*($C$11^2+A18^2)^(5/2))
D18: (F2) PR [W8] 3*$E$3*A18^3/(2*@PI*($D$11^2+A18^2)^(5/2))
E18: (F2) PR [W7] 3*$E$3/(2*@PI)*($E$11*A18^2/($E$11^2+A18^2)^(5/2))
F18: (F2) PR [W7] 3*$E$3/(2*@PI)*($F$11*A18^2/($F$11^2+A18^2)^(5/2))
G18: (F2) PR 3*$E$3/(2*@PI)*($G$11*A18^2/($G$11^2+A18^2)^(5/2))
A19: (F1) PR [W10] +A18+$E$5
B19: (F2) PR [W11] 3*$E$3*A19^3/(2*@PI*($B$11^2+A19^2)^(5/2))
C19: (F2) PR [W8] 3*$E$3*A19^3/(2*@PI*($C$11^2+A19^2)^(5/2))
D19: (F2) PR [W8] 3*$E$3*A19^3/(2*@PI*($D$11^2+A19^2)^(5/2))
E19: (F2) PR [W7] 3*$E$3/(2*@PI)*($E$11*A19^2/($E$11^2+A19^2)^(5/2))
F19: (F2) PR [W7] 3*$E$3/(2*@PI)*($F$11*A19^2/($F$11^2+A19^2)^(5/2))
G19: (F2) PR 3*$E$3/(2*@PI)*($G$11*A19^2/($G$11^2+A19^2)^(5/2))
A20: (F1) PR [W10] +A19+$E$5
B20: (F2) PR [W11] 3*$E$3*A20^3/(2*@PI*($B$11^2+A20^2)^(5/2))
C20: (F2) PR [W8] 3*$E$3*A20^3/(2*@PI*($C$11^2+A20^2)^(5/2))
D20: (F2) PR [W8] 3*$E$3*A20^3/(2*@PI*($D$11^2+A20^2)^(5/2))
E20: (F2) PR [W7] 3*$E$3/(2*@PI)*($E$11*A20^2/($E$11^2+A20^2)^(5/2))
F20: (F2) PR [W7] 3*$E$3/(2*@PI)*($F$11*A20^2/($F$11^2+A20^2)^(5/2))
G20: (F2) PR 3*$E$3/(2*@PI)*($G$11*A20^2/($G$11^2+A20^2)^(5/2))

3. FILE: WALL1

A1: PR 'Lateral Earth Pressure due to line loading
A2: PR \-
B2: PR \-
C2: PR \-
D2: PR \-
E2: PR \-
A3: PR 'This program calculates the horizontal pressure
A4: PR 'on rigid walls from a surface line load, using
A5: PR 'the modified Boussinesq equation.
A7: PR 'Maximum Wall Height (H) =
D7: U 5.1
E7: PR 'metres
A9: PR \-
B9: PR \-
C9: PR \-
D9: PR \-
E9: PR \-
C10: PR 'Q / unit length
A11: PR 'Line Loading
C11: PR 'Load 1
D11: PR 'Load 2
E11: PR 'Combined
A12: PR \-
B12: PR \-
C12: PR \-
D12: PR \-
E12: PR \+
A13: PR 'Magnitude kN
C13: U 55
D13: U 55
E13: PR +C13+D13
A14: PR 'Distance x(metres)
C14: U 2.1

D14: U 3.6
E14: PR *
A15: PR "Ratio m = x/H
C15: (F3) PR +C14/D7
D15: (F3) PR +D14/D7
E15: PR *
A16: PR 'Total thrust (kN)
C16: (F1) PR @IF(C15>0.4,2*C13/@PI/(C15^2+1),0.548*C13)
D16: (F1) PR @IF(D15>0.4,2*D13/@PI/(D15^2+1),0.548*D13)
E16: (F1) PR +C16+D16
A17: PR 'P(h) max (kPa)
C17: (F2) PR +C16/D7
D17: (F2) PR +D16/D7
E17: (F2) PR +C17+D17
A18: PR \-
B18: PR \-
C18: PR \-
D18: PR \-
E18: PR \+
A19: PR 'P(h) max = Equivalent Uniform Lateral Pressure
A21: PR \+
B21: PR \+
C21: PR \+
D21: PR \+
E21: PR \+
A22: PR ^Depth
B22: PR ^Ratio
C22: PR 'Horizontal pressure (kPa)
A23: PR ^z (m)
B23: PR ^n = z/H
C23: PR "Load 1
D23: PR ^Load 2
E23: PR "Combined
A24: PR \+
B24: PR \+
C24: PR \+
D24: PR \+
E24: PR \+
A25: (F1) U 0
B25: (F3) PR +A25/$D$7
C25: (F2)PR @IF($C$15>0.4,4*$C$13/@PI/$D$7*($C$15*$C$15*B25/($C$15^2+B25^2)^2),$C$1
3/$D$7*0.203*B25/(0.16+B25^2)^2)
D25: (F2) PR @IF($D$15>0.4,4*$D$13/@PI/$D$7*($D$15*$D$15*B25/($D$15^2+B25^2)^2),D
13/$D$7*0.203*B25/(0.16+B25^2)^2)
E25: (F2) PR +C25+D25
A26: (F1) PR +A25+0.5
B26: (F3) PR +A26/$D$7
C26: (F2) PR @IF($C$15>0.4,4*$C$13/@PI/$D$7*($C$15*$C$15*B26/($C$15^2+B26^2)^2),$C$1
3/$D$7*0.203*B26/(0.16+B26^2)^2)
D26: (F2) PR @IF($D$15>0.4,4*$D$13/@PI/$D$7*($D$15*$D$15*B26/($D$15^2+B26^2)^2),D
13/$D$7*0.203*B26/(0.16+B26^2)^2)
E26: (F2) PR +C26+D26
A27: (F1) PR +A26+0.5
B27: (F3) PR +A27/$D$7
C27: (F2) PR @IF($C$15>0.4,4*$C$13/@PI/$D$7*($C$15*$C$15*B27/($C$15^2+B27^2)^2),$C$1
3/$D$7*0.203*B27/(0.16+B27^2)^2)
D27: (F2) PR @IF($D$15>0.4,4*$D$13/@PI/$D$7*($D$15*$D$15*B27/($D$15^2+B27^2)^2),D

13/$D$7*0.203*B27/(0.16+B27^2)^2)

E27: (F2) PR +C27+D27

A28: (F1) PR +A27+0.5

B28: (F3) PR +A28/$D$7

C28: (F2) PR @IF($C$15>0.4,4*$C$13/@PI/$D$7*($C$15*$C$15*B28/($C$15^2+B28^2)^2),$C$1
3/$D$7*0.203*B28/(0.16+B28^2)^2)

D28: (F2) PR @IF($D$15>0.4,4*$D$13/@PI/$D$7*($D$15*$D$15*B28/($D$15^2+B28^2)^2),D
13/$D$7*0.203*B28/(0.16+B28^2)^2)

E28: (F2) PR +C28+D28

A29: (F1) PR +A28+0.5

B29: (F3) PR +A29/$D$7

C29: (F2) PR @IF($C$15>0.4,4*$C$13/@PI/$D$7*($C$15*$C$15*B29/($C$15^2+B29^2)^2),$C$1
3/$D$7*0.203*B29/(0.16+B29^2)^2)

D29: (F2) PR @IF($D$15>0.4,4*$D$13/@PI/$D$7*($D$15*$D$15*B29/($D$15^2+B29^2)^2),D
13/$D$7*0.203*B29/(0.16+B29^2)^2)

E29: (F2) PR +C29+D29

A30: (F1) PR +A29+0.5

B30: (F3) PR +A30/$D$7

C30: (F2) PR @IF($C$15>0.4,4*$C$13/@PI/$D$7*($C$15*$C$15*B30/($C$15^2+B30^2)^2),$C$1
3/$D$7*0.203*B30/(0.16+B30^2)^2)

D30: (F2) PR @IF($D$15>0.4,4*$D$13/@PI/$D$7*($D$15*$D$15*B30/($D$15^2+B30^2)^2),D
13/$D$7*0.203*B30/(0.16+B30^2)^2)

E30: (F2) PR +C30+D30

A31: (F1) PR +A30+0.5

B31: (F3) PR +A31/$D$7

C31: (F2) PR @IF($C$15>0.4,4*$C$13/@PI/$D$7*($C$15*$C$15*B31/($C$15^2+B31^2)^2),$C$1
3/$D$7*0.203*B31/(0.16+B31^2)^2)

D31: (F2) PR @IF($D$15>0.4,4*$D$13/@PI/$D$7*($D$15*$D$15*B31/($D$15^2+B31^2)^2),D
13/$D$7*0.203*B31/(0.16+B31^2)^2)

E31: (F2) PR +C31+D31

A32: (F1) PR +A31+0.5

B32: (F3) PR +A32/$D$7

C32: (F2) PR @IF($C$15>0.4,4*$C$13/@PI/$D$7*($C$15*$C$15*B32/($C$15^2+B32^2)^2),$C$1
3/$D$7*0.203*B32/(0.16+B32^2)^2)

D32: (F2) PR @IF($D$15>0.4,4*$D$13/@PI/$D$7*($D$15*$D$15*B32/($D$15^2+B32^2)^2),D
13/$D$7*0.203*B32/(0.16+B32^2)^2)

E32: (F2) PR +C32+D32

A33: (F1) PR +A32+0.5

B33: (F3) PR +A33/$D$7

C33: (F2) PR @IF($C$15>0.4,4*$C$13/@PI/$D$7*($C$15*$C$15*B33/($C$15^2+B33^2)^2),$C$1
3/$D$7*0.203*B33/(0.16+B33^2)^2)

D33: (F2) PR @IF($D$15>0.4,4*$D$13/@PI/$D$7*($D$15*$D$15*B33/($D$15^2+B33^2)^2),D
13/$D$7*0.203*B33/(0.16+B33^2)^2)

E33: (F2) PR +C33+D33

A34: (F1) PR +A33+0.5

B34: (F3) PR +A34/$D$7

C34: (F2) PR @IF($C$15>0.4,4*$C$13/@PI/$D$7*($C$15*$C$15*B34/($C$15^2+B34^2)^2),$C$1
3/$D$7*0.203*B34/(0.16+B34^2)^2)

D34: (F2) PR @IF($D$15>0.4,4*$D$13/@PI/$D$7*($D$15*$D$15*B34/($D$15^2+B34^2)^2),D
13/$D$7*0.203*B34/(0.16+B34^2)^2)

E34: (F2) PR +C34+D34

A35: (F1) PR +A34+0.5

B35: (F3) PR +A35/$D$7

C35: (F2) PR @IF($C$15>0.4,4*$C$13/@PI/$D$7*($C$15*$C$15*B35/($C$15^2+B35^2)^2),$C$1
3/$D$7*0.203*B35/(0.16+B35^2)^2)

D35: (F2) PR @IF($D$15>0.4,4*$D$13/@PI/$D$7*($D$15*$D$15*B35/($D$15^2+B35^2)^2),D
13/$D$7*0.203*B35/(0.16+B35^2)^2)

E35: (F2) PR +C35+D35
A36: (F1) PR +A35+0.5
B36: (F3) PR +A36/$D$7
C36: (F2) PR @IF($C$15>0.4,4*$C$13/@PI/$D$7*($C$15*$C$15*B36/($C$15^2+B36^2)^2),$C$13/$D$7*0.203*B36/(0.16+B36^2)^2)
D36: (F2) PR @IF($D$15>0.4,4*$D$13/@PI/$D$7*($D$15*$D$15*B36/($D$15^2+B36^2)^2),$D$13/$D$7*0.203*B36/(0.16+B36^2)^2)
E36: (F2) PR +C36+D36
A37: (F1) PR +A36+0.5
B37: (F3) PR +A37/$D$7
C37: (F2) PR @IF($C$15>0.4,4*$C$13/@PI/$D$7*($C$15*$C$15*B37/($C$15^2+B37^2)^2),$C$13/$D$7*0.203*B37/(0.16+B37^2)^2)
D37: (F2) PR @IF($D$15>0.4,4*$D$13/@PI/$D$7*($D$15*$D$15*B37/($D$15^2+B37^2)^2),$D$13/$D$7*0.203*B37/(0.16+B37^2)^2)
E37: (F2) PR +C37+D37
A38: (F1) PR +A37+0.5
B38: (F3) PR +A38/$D$7
C38: (F2) PR @IF($C$15>0.4,4*$C$13/@PI/$D$7*($C$15*$C$15*B38/($C$15^2+B38^2)^2),$C$13/$D$7*0.203*B38/(0.16+B38^2)^2)
D38: (F2) PR @IF($D$15>0.4,4*$D$13/@PI/$D$7*($D$15*$D$15*B38/($D$15^2+B38^2)^2),$D$13/$D$7*0.203*B38/(0.16+B38^2)^2)
E38: (F2) PR +C38+D38

4. FILE: WALL2

I1: PR 'Lateral Earth Pressure due to compaction effort
I2: PR \-
J2: PR \-
K2: PR \-
L2: PR \-
M2: PR [W12] \-
I3: PR 'This program calculates the horizontal pressure on rigid walls
I4: PR 'from a surface load, with compaction induced effects
I6: PR 'Geometry & Soil Properties
N6: PR 'Imposed Load
I7: PR \-
J7: PR \-
K7: PR \-
N7: PR \-
O7: PR \-
P7: PR \-
Q7: PR \-
I8: PR 'Wall Height (H)
K8: U 5.1
L8: PR 'metres
N8: PR 'Dead weight
P8: U 125
Q8: PR 'kN/m
I9: PR 'Distance from wall
K9: U 3.85
L9: PR 'metres
N9: PR 'Centrifugal force
P9: U 0
Q9: PR 'kN/m
N10: PR 'Width of load

P10: U 2.5
Q10: PR 'metres
I11: PR 'Unit weight (kN/cu m)
L11: U 18
N11: PR 'Length of load
P11: U 100
Q11: PR 'metres
I12: PR 'Compact unit wght (kN/cu m)
L12: U 19
N12: PR 'P(roller load)
Q12: (F1) PR (P8+P9)/P10
R12: PR 'kN/sq m
I13: PR 'Minimum unit wght (kN/cu m)
L13: U 17
N13: PR 'Critical depth Z(c)
Q13: (F3) PR +$L$14*@SQRT(2*$Q$12/@PI/$L$11)
R13: PR 'metres
I14: PR 'Active earth pressure Ka
L14: U 0.3
N14: PR 'Depth of influence d
Q14: (F3) PR @SQRT(2*$Q$12/@PI/$L$11)/$L$14
R14: PR 'metres
I15: PR 'At Rest earth pressure Ko
L15: U 0.5
N15: PR 'Max horizontal pressure
Q15: (F2) PR @SQRT(2*Q12*L11/@PI)*P11/(K9+P11)
R15: PR 'kPa
I16: PR 'Compacted at rest press Koc
L16: PR +L15+(L12-L13)*0.025
N16: PR 'sustained after compaction
I17: PR \+
J17: PR \+
K17: PR \+
L17: PR \+
M17: PR [W12] \+
N17: PR \+
O17: PR \+
P17: PR \+
Q17: PR \+
M18: PR [W12] 'Compaction
N18: PR "Modified
K19: PR 'Active
L19: PR 'At Rest
M19: PR [W12] ^Induced
N19: PR 'At Rest
I20: PR 'Total thrust (kN)
K20: (F2) PR +$L$14*$L$11*$K$8^2/2
L20: (F2) PR +$L$15*$L$11*$K$8^2/2
M20: (F2) PR [W12] @IF($K$8<$Q$13,$K$8/$Q$13*$Q$15*$K$8/2,@IF($K$8<$Q$14,$Q$15*($K$8-$Q$13)+$Q$15*$Q$13/2,$Q$15*($K$8-$Q$13)+($Q$15*$Q$13/2)+($K$8-$Q$14)*$L$14*$L$11*($K$8+$Q$14)/2))
N20: (F2) PR +$L$16*$L$11*$K$8^2/2
I21: PR \+
J21: PR \+
K21: PR \+
L21: PR \+
M21: PR [W12] \+

N21: PR \+
K22: PR 'Horizontal pressure (kPa)
I23: PR 'Depth z (metres)
K23: PR 'Active
L23: PR 'At Rest
M23: PR [W12] 'Induced
N23: PR 'Modified
I24: PR \+
J24: PR \+
K24: PR \+
L24: PR \+
M24: PR [W12] \+
N24: PR \+
I25: (F1) U 0.00001
K25: (F2) PR +$L$14*$L$11*I25
L25: (F2) PR +$L$15*$L$11*I25
M25: (F2) PR [W12] @IF(I25<$Q$13,I25/$Q$13*$Q$15,@IF(I25<$Q$14,$Q$15,K25))
N25: (F2) PR +$L$16*$L$11*I25
I26: (F1) PR +I25+0.5
K26: (F2) PR +$L$14*$L$11*I26
L26: (F2) PR +$L$15*$L$11*I26
M26: (F2) PR [W12] @IF(I26<$Q$13,I26/$Q$13*$Q$15,@IF(I26<$Q$14,$Q$15,K26))
N26: (F2) PR +$L$16*$L$11*I26
I27: (F1) PR +I26+0.5
K27: (F2) PR +$L$14*$L$11*I27
L27: (F2) PR +$L$15*$L$11*I27
M27: (F2) PR [W12] @IF(I27<$Q$13,I27/$Q$13*$Q$15,@IF(I27<$Q$14,$Q$15,K27))
N27: (F2) PR +$L$16*$L$11*I27
I28: (F1) PR +I27+0.5
K28: (F2) PR +$L$14*$L$11*I28
L28: (F2) PR +$L$15*$L$11*I28
M28: (F2) PR [W12] @IF(I28<$Q$13,I28/$Q$13*$Q$15,@IF(I28<$Q$14,$Q$15,K28))
N28: (F2) PR +$L$16*$L$11*I28
I29: (F1) PR +I28+0.5
K29: (F2) PR +$L$14*$L$11*I29
L29: (F2) PR +$L$15*$L$11*I29
M29: (F2) PR [W12] @IF(I29<$Q$13,I29/$Q$13*$Q$15,@IF(I29<$Q$14,$Q$15,K29))
N29: (F2) PR +$L$16*$L$11*I29
I30: (F1) PR +I29+0.5
K30: (F2) PR +$L$14*$L$11*I30
L30: (F2) PR +$L$15*$L$11*I30
M30: (F2) PR [W12] @IF(I30<$Q$13,I30/$Q$13*$Q$15,@IF(I30<$Q$14,$Q$15,K30))
N30: (F2) PR +$L$16*$L$11*I30
I31: (F1) PR +I30+0.5
K31: (F2) PR +$L$14*$L$11*I31
L31: (F2) PR +$L$15*$L$11*I31
M31: (F2) PR [W12] @IF(I31<$Q$13,I31/$Q$13*$Q$15,@IF(I31<$Q$14,$Q$15,K31))
N31: (F2) PR +$L$16*$L$11*I31
I32: (F1) PR +I31+0.5
K32: (F2) PR +$L$14*$L$11*I32
L32: (F2) PR +$L$15*$L$11*I32
M32: (F2) PR [W12] @IF(I32<$Q$13,I32/$Q$13*$Q$15,@IF(I32<$Q$14,$Q$15,K32))
N32: (F2) PR +$L$16*$L$11*I32
I33: (F1) PR +I32+0.5
K33: (F2) PR +$L$14*$L$11*I33
L33: (F2) PR +$L$15*$L$11*I33
M33: (F2) PR [W12] @IF(I33<$Q$13,I33/$Q$13*$Q$15,@IF(I33<$Q$14,$Q$15,K33))

N33: (F2) PR +$L$16*$L$11*I33
I34: (F1) PR +I33+0.5
K34: (F2) PR +$L$14*$L$11*I34
L34: (F2) PR +$L$15*$L$11*I34
M34: (F2) PR [W12] @IF(I34<$Q$13,I34/$Q$13*$Q$15,@IF(I34<$Q$14,$Q$15,K34))
N34: (F2) PR +$L$16*$L$11*I34
I35: (F1) PR +I34+0.5
K35: (F2) PR +$L$14*$L$11*I35
L35: (F2) PR +$L$15*$L$11*I35
M35: (F2) PR [W12] @IF(I35<$Q$13,I35/$Q$13*$Q$15,@IF(I35<$Q$14,$Q$15,K35))
N35: (F2) PR +$L$16*$L$11*I35
I36: (F1) PR +I35+0.5
K36: (F2) PR +$L$14*$L$11*I36
L36: (F2) PR +$L$15*$L$11*I36
M36: (F2) PR [W12] @IF(I36<$Q$13,I36/$Q$13*$Q$15,@IF(I36<$Q$14,$Q$15,K36))
N36: (F2) PR +$L$16*$L$11*I36
I37: (F1) PR +I36+0.5
K37: (F2) PR +$L$14*$L$11*I37
L37: (F2) PR +$L$15*$L$11*I37
M37: (F2) PR [W12] @IF(I37<$Q$13,I37/$Q$13*$Q$15,@IF(I37<$Q$14,$Q$15,K37))
N37: (F2) PR +$L$16*$L$11*I37
I38: (F1) PR +I37+0.5
K38: (F2) PR +$L$14*$L$11*I38
L38: (F2) PR +$L$15*$L$11*I38
M38: (F2) PR [W12] @IF(I38<$Q$13,I38/$Q$13*$Q$15,@IF(I38<$Q$14,$Q$15,K38))
N38: (F2) PR +$L$16*$L$11*I38
I42: PR 'Height of
K42: PR ^Total Thrust (kN)
I43: PR 'Wall (m)
K43: PR 'Active
L43: PR 'At Rest
M43: PR [W12] 'Induced
N43: PR 'Modified
I44: PR \+
J44: PR \+
K44: PR \+
L44: PR \+
M44: PR [W12] \+
N44: PR \+
I45: U 0
K45: (F2) PR +$L$14*$L$11*I45^2/2
L45: (F2) PR +$L$15*$L$11*I45^2/2
M45: (F2) PR [W12] @IF(I45<$Q$13,I45/$Q$13*$Q$15*I45/2,@IF(I45<$Q$14,$Q$15*(I45-Q
13)+$Q$15*$Q$13/2,$Q$15*(I45-$Q$13))+($Q$15*$Q$13/2)+(I45-$Q$14)*$L$14*$L$11*(I45+$Q$14)
/2))
N45: (F2) PR +$L$16*$L$11*I45^2/2
I46: (F1) PR +I45+1
K46: (F2) PR +$L$14*$L$11*I46^2/2
L46: (F2) PR +$L$15*$L$11*I46^2/2
M46: (F2) PR [W12] @IF(I46<$Q$13,I46/$Q$13*$Q$15*I46/2,@IF(I46<$Q$14,$Q$15*(I46-Q
13)+$Q$15*$Q$13/2,$Q$15*(I46-$Q$13))+($Q$15*$Q$13/2)+(I46-$Q$14)*$L$14*$L$11*(I46+$Q$14)
/2))
N46: (F2) PR +$L$16*$L$11*I46^2/2
I47: (F1) PR +I46+1
K47: (F2) PR +$L$14*$L$11*I47^2/2
L47: (F2) PR +$L$15*$L$11*I47^2/2
M47: (F2) PR [W12] @IF(I47<$Q$13,I47/$Q$13*$Q$15*I47/2,@IF(I47<$Q$14,$Q$15*(I47-Q

13)+$Q$15*$Q$13/2,$Q$15*(I47-$Q$13)+($Q$15*$Q$13/2)+(I47-$Q$14)*$L$14*$L$11*(I47+$Q$14)/2))

N47: (F2) PR +$L$16*$L$11*I47^2/2
I48: (F1) PR +I47+1
K48: (F2) PR +$L$14*$L$11*I48^2/2
L48: (F2) PR +$L$15*$L$11*I48^2/2
M48: (F2) PR [W12] @IF(I48<$Q$13,I48/$Q$13*$Q$15*I48/2,@IF(I48<$Q$14,$Q$15*(I48-$Q$13)+$Q$15*$Q$13/2,$Q$15*(I48-$Q$13)+($Q$15*$Q$13/2)+(I48-$Q$14)*$L$14*$L$11*(I48+$Q$14)/2))

N48: (F2) PR +$L$16*$L$11*I48^2/2
I49: (F1) PR +I48+1
K49: (F2) PR +$L$14*$L$11*I49^2/2
L49: (F2) PR +$L$15*$L$11*I49^2/2
M49: (F2) PR [W12] @IF(I49<$Q$13,I49/$Q$13*$Q$15*I49/2,@IF(I49<$Q$14,$Q$15*(I49-$Q$13)+$Q$15*$Q$13/2,$Q$15*(I49-$Q$13)+($Q$15*$Q$13/2)+(I49-$Q$14)*$L$14*$L$11*(I49+$Q$14)/2))

N49: (F2) PR +$L$16*$L$11*I49^2/2
I50: (F1) PR +I49+1
K50: (F2) PR +$L$14*$L$11*I50^2/2
L50: (F2) PR +$L$15*$L$11*I50^2/2
M50: (F2) PR [W12] @IF(I50<$Q$13,I50/$Q$13*$Q$15*I50/2,@IF(I50<$Q$14,$Q$15*(I50-$Q$13)+$Q$15*$Q$13/2,$Q$15*(I50-$Q$13)+($Q$15*$Q$13/2)+(I50-$Q$14)*$L$14*$L$11*(I50+$Q$14)/2))

N50: (F2) PR +$L$16*$L$11*I50^2/2
I51: (F1) PR +I50+1
K51: (F2) PR +$L$14*$L$11*I51^2/2
L51: (F2) PR +$L$15*$L$11*I51^2/2
M51: (F2) PR [W12] @IF(I51<$Q$13,I51/$Q$13*$Q$15*I51/2,@IF(I51<$Q$14,$Q$15*(I51-$Q$13)+$Q$15*$Q$13/2,$Q$15*(I51-$Q$13)+($Q$15*$Q$13/2)+(I51-$Q$14)*$L$14*$L$11*(I51+$Q$14)/2))

N51: (F2) PR +$L$16*$L$11*I51^2/2

5. FILE: SLOPESOI

A1: PR [W11] 'Stability of a Planar Slope
D1: PR 'JOB NO :
E1: U 'Brisbane
A2: PR [W11] _
B2: PR _
C2: PR _
D2: PR \+
E2: PR \+
F2: PR [W10] \+
A4: PR [W11] 'SLOPE ANGLE
B4: PR "degrees
C4: PR "Radians
E4: PR 'GEOMETERY OF WEDGE
A5: PR [W11] \-
E5: PR \-
F5: PR [W10] \-
A6: PR [W11] 'Face
B6: U 90
C6: PR @PI*B6/180
E6: PR 'Slope Height - H
G6: U 7

```
H6: PR [W11] 'metres
A7: PR [W11] 'Plane
B7: U 70
C7: PR @PI*B7/180
E7: PR 'Tension crack - z
G7: U 2.3
H7: PR [W11] 'metres
E8: PR 'Water Depth  - z(w)
G8: U 0
H8: PR [W11] 'metres
A9: PR [W11] 'EXTERNAL LOADING
E9: PR 'Tension crack - z(o)
G9: U 0
H9: PR [W11] 'metres
A10: PR [W11] \-
B10: PR \-
E10: PR 'z(o) = 0 for crack at crown of slope
A11: PR [W11] 'Surcharge
C11: U 55
D11: PR 'kPa
E11: PR 'z(o) > 0 for crack on slope face
A12: PR [W11] 'Point Load   P
C12: U 400
D12: PR 'kN
A13: PR [W11] 'Dist. from edge Xp
C13: U 1
D13: PR 'metres
E13: PR 'Total crown - X
G13: PR +G6/@TAN(C7)-G6/@TAN(C6)
H13: PR [W11] 'metres
A14: PR [W11] 'Lateral Load  L
C14: U 32
D14: PR 'kN
E14: PR 'Edge of Crown - D
G14: PR +G7/@TAN(C7)
H14: PR [W11] 'metres
A15: PR [W11] 'Buttress Load  B
C15: U 0
D15: PR 'kN
E15: PR 'Plane length - A
G15: PR (G6-G7)/@SIN(C7)
H15: PR [W11] 'metres
A16: PR [W11] 'Bolt Tension  T
C16: U 1380
D16: PR 'kN
A17: PR [W11] 'at Normal Angle of
C17: U 5
D17: PR 'degrees to failure plane
A18: PR [W11] 'at Normal Angle of
C18: PR +C17-(B6-B7)
D18: PR 'degrees to slope face
A20: PR [W11] 'GROUNDWATER CONDITIONS
D20: U 'c
A21: PR [W11] \-
B21: PR \-
D21: PR "?????????
A22: PR [W11] 'Dry Slope
```

B22: PR ': y
C22: PR 'Water conditions (a,b or c)
B23: PR "a :
C23: PR 'In Tension Crack Only
B24: PR "b :
C24: PR 'In Tension crack & sliding surface
B25: PR "c :
C25: PR 'Saturated slope with heavy recharge
A28: PR [W11] 'SOIL PROPERTIES
C28: PR 'Average
D28: PR 'Minimum
E28: PR 'Maximum
A29: PR [W11] \-
B29: PR \-
C29: PR \-
D29: PR \-
E29: PR \-
A30: PR [W11] 'Cohesion (kPa)
C30: U 10
D30: U 5
E30: U 15
A31: PR [W11] 'Angle of friction
C31: U 25
D31: U 20
E31: U 30
A32: PR [W11] 'Unit Wght (kN/cu m)
C32: U 24
D32: PR +C32
E32: PR +C32
A33: PR [W11] \-
B33: PR \-
C33: PR \-
D33: PR \-
E33: PR \-
A35: PR [W11] 'FORCES (kN / m)
F35: PR [W10] 'TOTAL LOAD
A36: PR [W11] \+
B36: PR \+
C36: PR \+
D36: PR \+
F36: PR [W10] \+
G36: PR \+
A37: PR [W11] 'Weight of Sliding wedge
D37: (F2) PR 0.5*C32*(G6*G13-G14*G7+G9*(G14-G13))
F37: PR [W10] 'Vertical
G37: (F2) PR @IF(C13<G13,+D37+C12+C11*G13,+D37+C11*G13)
A38: PR [W11] 'Horizontal water force
D38: (F2) PR @IF(D20="y",0,@IF(D20="a",0.5*9.81*G8*G8,@IF(D20="b",0.5*9.81*G8*G8,0.5*
9.81*G7*G7)))
F38: PR [W10] 'Horizontal
G38: (F2) PR +D38+C14-C15
A39: PR [W11] 'Uplift Water force
D39: (F2) PR @IF(D20="y",0,@IF(D20="a",0,@IF(D20="b",0.5*9.81*G8*G15,0.5*9.81*G7*G1
5)))
F39: PR [W10] 'Uplift
G39: (F2) PR +D39
A41: PR [W11] 'ACTIVATING Forces (kN) =

D41: (F2) PR ($G$37*@SIN($C$7)+$G$38*@COS($C$7)-$C$16*@SIN($C$17*@PI/180))
E41: PR 'and independent of strength paramaters
A43: PR [W11] 'Calculation of Slope Safety factor variation with soil properties
A44: PR [W11] \+
B44: PR \+
C44: PR \+
D44: PR \+
E44: PR \+
F44: PR [W10] \+
B45: PR 'R E S I S T I N G
D45: PR 'F O R C E S (kN)
A46: PR [W11] 'Cohesion
C46: PR 'Angle of friction (degrees)
A47: PR [W11] "(kPa)
B47: (F1) PR +$D$31
C47: (F1) PR ($D$31+$C$31)/2
D47: (F1) PR +C31
E47: (F1) PR (C31+E31)/2
F47: (F1) PR [W10] +E31
A48: PR [W11] \+
B48: PR \+
C48: PR \+
D48: PR \+
E48: PR \+
F48: PR [W10] \+
A49: (F1) PR [W11] +D30
B49: (F2) PR ($A$49*$G$15+($G$37*@COS($C$7)-$G$39-$G$38*@SIN($C$7)+$C$16*@COS($
C$17*@PI/180))*@TAN(B47*@PI/180))
C49: (F2) PR (A49*$G$15+($G$37*@COS($C$7)-$G$39-$G$38*@SIN($C$7)+$C$16*@COS($
C$17*@PI/180))*@TAN($C$47*@PI/180))
D49: (F2) PR (A49*$G$15+($G$37*@COS($C$7)-$G$39-$G$38*@SIN($C$7)+$C$16*@COS($
C$17*@PI/180))*@TAN($D$47*@PI/180))
E49: (F2) PR (A49*$G$15+($G$37*@COS($C$7)-$G$39-$G$38*@SIN($C$7)+$C$16*@COS($
C$17*@PI/180))*@TAN($E$47*@PI/180))
F49: (F2) PR [W10] (A49*$G$15+($G$37*@COS($C$7)-$G$39-$G$38*@SIN($C$7)+$C$16*@
COS($C$17*@PI/180))*@TAN($F$47*@PI/180))
G49: (H) PR +$D$41
A50: (F1) PR [W11] (+C30+D30)/2
B50: (F2) PR (A50*$G$15+($G$37*@COS($C$7)-$G$39-$G$38*@SIN($C$7)+$C$16*@COS($
C$17*@PI/180))*@TAN($B$47*@PI/180))
C50: (F2) PR (A50*$G$15+($G$37*@COS($C$7)-$G$39-$G$38*@SIN($C$7)+$C$16*@COS($
C$17*@PI/180))*@TAN($C$47*@PI/180))
D50: (F2) PR (A50*$G$15+($G$37*@COS($C$7)-$G$39-$G$38*@SIN($C$7)+$C$16*@COS($
C$17*@PI/180))*@TAN($D$47*@PI/180))
E50: (F2) PR (A50*$G$15+($G$37*@COS($C$7)-$G$39-$G$38*@SIN($C$7)+$C$16*@COS($
C$17*@PI/180))*@TAN($E$47*@PI/180))
F50: (F2) PR [W10] (A50*$G$15+($G$37*@COS($C$7)-$G$39-$G$38*@SIN($C$7)+$C$16*@
COS($C$17*@PI/180))*@TAN($F$47*@PI/180))
G50: (H) PR +$D$41
A51: (F1) PR [W11] +C30
B51: (F2) PR (A51*$G$15+($G$37*@COS($C$7)-$G$39-$G$38*@SIN($C$7)+$C$16*@COS($
C$17*@PI/180))*@TAN($B$47*@PI/180))
C51: (F2) PR (A51*$G$15+($G$37*@COS($C$7)-$G$39-$G$38*@SIN($C$7)+$C$16*@COS($
C$17*@PI/180))*@TAN($C$47*@PI/180))
D51: (F2) PR (A51*$G$15+($G$37*@COS($C$7)-$G$39-$G$38*@SIN($C$7)+$C$16*@COS($
C$17*@PI/180))*@TAN($D$47*@PI/180))
E51: (F2) PR (A51*$G$15+($G$37*@COS($C$7)-$G$39-$G$38*@SIN($C$7)+$C$16*@COS($

C$17*@PI/180))*@TAN($E$47*@PI/180))

F51: (F2) PR [W10] (A51*$G$15+($G$37*@COS($C$7)-$G$39-$G$38*@SIN($C$7)+$C$16*@
COS($C$17*@PI/180))*@TAN($F$47*@PI/180))

G51: (H) PR +$D$41

A52: (F1) PR [W11] (+C30+E30)/2

B52: (F2) PR (A52*$G$15+($G$37*@COS($C$7)-$G$39-$G$38*@SIN($C$7)+$C$16*@COS($
C$17*@PI/180))*@TAN($B$47*@PI/180))

C52: (F2) PR (A52*$G$15+($G$37*@COS($C$7)-$G$39-$G$38*@SIN($C$7)+$C$16*@COS($
C$17*@PI/180))*@TAN($C$47*@PI/180))

D52: (F2) PR (A52*$G$15+($G$37*@COS($C$7)-$G$39-$G$38*@SIN($C$7)+$C$16*@COS($
C$17*@PI/180))*@TAN($D$47*@PI/180))

E52: (F2) PR (A52*$G$15+($G$37*@COS($C$7)-$G$39-$G$38*@SIN($C$7)+$C$16*@COS($
C$17*@PI/180))*@TAN($E$47*@PI/180))

F52: (F2) PR [W10] (A52*$G$15+($G$37*@COS($C$7)-$G$39-$G$38*@SIN($C$7)+$C$16*@
COS($C$17*@PI/180))*@TAN($F$47*@PI/180))

G52: (H) PR +$D$41

A53: (F1) PR [W11] +E30

B53: (F2) PR (A53*$G$15+($G$37*@COS($C$7)-$G$39-$G$38*@SIN($C$7)+$C$16*@COS($
C$17*@PI/180))*@TAN($B$47*@PI/180))

C53: (F2) PR (A53*$G$15+($G$37*@COS($C$7)-$G$39-$G$38*@SIN($C$7)+$C$16*@COS($
C$17*@PI/180))*@TAN($C$47*@PI/180))

D53: (F2) PR (A53*$G$15+($G$37*@COS($C$7)-$G$39-$G$38*@SIN($C$7)+$C$16*@COS($
C$17*@PI/180))*@TAN($D$47*@PI/180))

E53: (F2) PR (A53*$G$15+($G$37*@COS($C$7)-$G$39-$G$38*@SIN($C$7)+$C$16*@COS($
C$17*@PI/180))*@TAN($E$47*@PI/180))

F53: (F2) PR [W10] (A53*$G$15+($G$37*@COS($C$7)-$G$39-$G$38*@SIN($C$7)+$C$16*@
COS($C$17*@PI/180))*@TAN($F$47*@PI/180))

G53: (H) PR +$D$41

A55: PR [W11] \+

B55: PR \+

C55: PR \+

D55: PR \+

E55: PR \+

F55: PR [W10] \+

C56: PR 'F A C T O R   O F   S A F E T Y

A57: PR [W11] 'Cohesion

C57: PR 'Angle of friction (degrees)

A58: PR [W11] "(kPa)

B58: (F1) PR +B47

C58: (F1) PR +C47

D58: (F1) PR +D47

E58: (F1) PR +E47

F58: (F1) PR [W10] +F47

A59: PR [W11] \+

B59: PR \+

C59: PR \+

D59: PR \+

E59: PR \+

F59: PR [W10] \+

A60: PR [W11] +A49

B60: (F2) PR +B49/$D$41

C60: (F2) PR +C49/$D$41

D60: (F2) PR +D49/$D$41

E60: (F2) PR +E49/$D$41

F60: (F2) PR [W10] +F49/$D$41

G60: (H) PR ($F$64-$B$60)/5+$F$64

A61: PR [W11] +A50

B61: (F2) PR +B50/$D$41
C61: (F2) PR +C50/$D$41
D61: (F2) PR +D50/$D$41
E61: (F2) PR +E50/$D$41
F61: (F2) PR [W10] +F50/$D$41
G61: (H) PR ($F$64-$B$60)/5+$F$64
A62: PR [W11] +A51
B62: (F2) PR +B51/$D$41
C62: (F2) PR +C51/$D$41
D62: (F2) PR +D51/$D$41
E62: (F2) PR +E51/$D$41
F62: (F2) PR [W10] +F51/$D$41
G62: (H) PR ($F$64-$B$60)/5+$F$64
A63: PR [W11] +A52
B63: (F2) PR +B52/$D$41
C63: (F2) PR +C52/$D$41
D63: (F2) PR +D52/$D$41
E63: (F2) PR +E52/$D$41
F63: (F2) PR [W10] +F52/$D$41
G63: (H) PR ($F$64-$B$60)/5+$F$64
A64: PR [W11] +A53
B64: (F2) PR +B53/$D$41
C64: (F2) PR +C53/$D$41
D64: (F2) PR +D53/$D$41
E64: (F2) PR +E53/$D$41
F64: (F2) PR [W10] +F53/$D$41
G64: (H) PR ($F$64-$B$60)/5+$F$64
A65: PR [W11] \+
B65: PR \+
C65: PR \+
D65: PR \+
E65: PR \+
F65: PR [W10] \+
A66: PR [W11] 'GRAPH AIDS
A67: PR [W11] @IF(D20="y",A21,@IF(D20="a",C23,@IF(D20="b",C24,C25)))
C67: PR 'Slope face at (degs)
D67: PR +B6
C68: PR 'ACTIVATING FORCE is independent of the strength paramaters

6. FILE: SLOPEGRO

A1: PR 'Stability of a Planar Slope
D1: PR 'JOB NO :
E1: U 'Brisbane
A2: PR _
B2: PR _
C2: PR _
D2: PR \+
E2: PR \+
A4: PR 'S L O P E
B4: PR " A N G L E
E4: PR 'GROUNDWATER CONDITIONS
A5: PR \-
B5: PR \-
C5: PR \-

E5: PR \-
F5: PR [W9] \-
G5: PR [W9] \-
A6: PR ^OF
B6: PR "(degrees)
C6: PR "(Radians)
D6: PR "d:
E6: PR 'Dry Slope
F6: PR [W9] ' Water conditions (a,b,c)
A7: PR 'Face
B7: U 90
C7: PR @PI*B7/180
D7: PR "a:
E7: PR 'In Tension Crack Only
A8: PR 'Plane avg
B8: U 70
C8: PR @PI*B8/180
D8: PR "b:
E8: PR 'In Tension crack & sliding surface
A9: PR 'Min. Plane
B9: U 60
C9: PR @PI*B9/180
D9: PR "c:
E9: PR 'Saturated slope with heavy recharge
A10: PR 'Max. Plane
B10: U 75
C10: PR @PI*B10/180
E11: PR 'GEOMETERY OF WEDGE
A12: PR 'SOIL PROPERTIES
E12: PR \-
F12: PR [W9] \-
G12: PR [W9] \-
H12: PR [W11] \-
A13: PR \-
B13: PR \-
C13: PR \-
E13: PR 'Slope Height - H
G13: U [W9] 7
H13: PR [W11] 'metres
A14: PR 'Cohesion (kPa)
C14: U 10
E14: PR 'Tension crack - z
G14: U [W9] 2.3
H14: PR [W11] 'metres
A15: PR 'Angle of friction
C15: U 25
E15: PR 'Water Depth - z(w)
G15: U [W9] 2
H15: PR [W11] 'metres
A16: PR 'Unit Wght(kN/cu m)
C16: U 24
E16: PR 'Tension crack- z(o)
G16: U [W9] 0
H16: PR [W11] 'below crown
A18: PR 'EXTERNAL LOADING
E18: PR 'z(o) = 0 for crack at crown of slope
A19: PR \-

B19: PR \-
C19: PR \-
E19: PR 'z(o) > 0 for crack on slope face
A20: PR 'Surcharge
C20: U 55
D20: PR 'kPa
A21: PR 'Point Load   P
C21: U 400
D21: PR 'kN
A22: PR 'Dist. from edge Xp
C22: U 1
D22: PR 'metres
E22: PR 'Total crown - X
G22: PR [W9] +G13/@TAN(C8)-G13/@TAN(C7)
H22: PR [W11] 'metres
A23: PR 'Lateral Load L
C23: U 32
D23: PR 'kN
E23: PR 'Edge of Crown - D
G23: PR [W9] +G14/@TAN(B8*@PI/180)
H23: PR [W11] 'metres
A24: PR 'Buttress   B
C24: U 0
D24: PR 'kN
E24: PR 'Plane length - A
G24: PR [W9] (G13-G14)/@SIN(C8)
H24: PR [W11] 'metres
A25: PR 'Bolt Tension T
C25: U 1380
D25: PR 'kN
A26: PR 'at Normal Angle of
C26: U 5
D26: PR 'degrees to failure plane
A27: PR 'at Normal Angle of
C27: PR +C26-(B7-B8)
D27: PR 'degrees to slope face
A28: PR 'Weight of Sliding wedge
D28: PR 0.5*C16*(G13*G22-G23*G14+G16*(G23-G22))
E28: PR 'kN/m
C31: PR 'G R O U N D W A T E R   C O N D I T I O N S
A32: PR 'Water Forces
C32: PR "A
D32: PR "B
E32: PR "C
F32: PR [W9] "D
A33: PR \-
B33: PR \-
C33: PR \-
D33: PR \-
E33: PR \-
F33: PR [W9] \-
G33: PR [W9] \-
A34: PR 'Horizontal (H)
C34: (F2) PR 0.5*9.81*$G$15^2
D34: (F2) PR 0.5*9.81*$G$15^2
E34: (F2) PR 0.5*9.81*$G$14^2
F34: (F2) PR [W9] 0

A35: PR 'Uplift (U)
C35: (F2) PR 0
D35: (F2) PR 0.5*9.81*$G$15*$G$24
E35: (F2) PR 0.5*9.81*$G$14*$G$24
F35: (F2) PR [W9] 0
A37: PR 'TOTAL LOAD (kN/m)
C37: PR "A
D37: PR "B
E37: PR "C
F37: PR [W9] "D
A38: PR \-
B38: PR \-
C38: PR \-
D38: PR \-
E38: PR \-
F38: PR [W9] \-
G38: PR [W9] \-
A39: PR 'Horizontal
C39: (F2) PR +$C$23+C34-$C$24
D39: (F2) PR +$C$23+D34-$C$24
E39: (F2) PR +$C$23+E34-$C$24
F39: (F2) PR [W9] +$C$23+F34-$C$24
A40: PR 'Uplift
C40: (F2) PR +C35
D40: (F2) PR +D35
E40: (F2) PR +E35
F40: (F2) PR [W9] +F35
A41: PR 'Vertical load (kN/m) =
C41: (F2) PR @IF(C22<G22,+D28+C21+C20*G22,+D28+C20*G22)
D41: PR 'and independent of water conditions
A44: PR 'Variation of Slope Safety factor with goundwater conditions and slope angle
A45: PR \+
B45: PR \+
C45: PR \+
D45: PR \+
E45: PR \+
F45: PR [W9] \+
G45: PR [W9] \+
A46: PR 'RADIAN
B46: PR 'G R O U N D W A T E R   C O N D I T I O N S
G46: PR [W9] 'DEGREE
A47: PR 'Plane
B47: PR "A
C47: PR "B
D47: PR "C
E47: PR "D
G47: PR [W9] 'Plane
A48: PR 'Angle
B48: PR 'R E S I S T I N G   F O R C E S (kN/m)
G48: PR [W9] 'Angle
A49: PR \+
B49: PR \+
C49: PR \+
D49: PR \+
E49: PR \+
F49: PR [W9] \+
G49: PR [W9] \+

A50: (F3) PR +C9

B50: (F2) PR ($C$14*$G$24+($C$41*@COS($A$50)-$C$40-$C$39*@SIN($A$50)+$C$25*@COS
($C$26*@PI/180))*@TAN($C$15*@PI/180))

C50: (F2) PR ($C$14*$G$24+($C$41*@COS(A50)-$D$40-$D$39*@SIN(A50)+$C$25*@COS
($C$26*@PI/180))*@TAN($C$15*@PI/180))

D50: (F2) PR ($C$14*$G$24+($C$41*@COS(A50)-$E$40-$E$39*@SIN(A50)+$C$25*@COS
($C$26*@PI/180))*@TAN($C$15*@PI/180))

E50: (F2) PR ($C$14*$G$24+($C$41*@COS(A50)-$F$40-$F$39*@SIN(A50)+$C$25*@COS
($C$26*@PI/180))*@TAN($C$15*@PI/180))

G50: PR [W9] +A50/@PI*180

A51: (F3) PR (+C8+C9)/2

B51: (F2) PR ($C$14*$G$24+($C$41*@COS(A51)-$C$40-$C$39*@SIN(A51)+$C$25*@COS
($C$26*@PI/180))*@TAN($C$15*@PI/180))

C51: (F2) PR ($C$14*$G$24+($C$41*@COS(A51)-$D$40-$D$39*@SIN(A51)+$C$25*@COS
($C$26*@PI/180))*@TAN($C$15*@PI/180))

D51: (F2) PR ($C$14*$G$24+($C$41*@COS(A51)-$E$40-$E$39*@SIN(A51)+$C$25*@COS
($C$26*@PI/180))*@TAN($C$15*@PI/180))

E51: (F2) PR ($C$14*$G$24+($C$41*@COS(A51)-$F$40-$F$39*@SIN(A51)+$C$25*@COS
($C$26*@PI/180))*@TAN($C$15*@PI/180))

G51: PR [W9] +A51/@PI*180

A52: (F3) PR +C8

B52: (F2) PR ($C$14*$G$24+($C$41*@COS(A52)-$C$40-$C$39*@SIN(A52)+$C$25*@COS
($C$26*@PI/180))*@TAN($C$15*@PI/180))

C52: (F2) PR ($C$14*$G$24+($C$41*@COS(A52)-$D$40-$D$39*@SIN(A52)+$C$25*@COS
($C$26*@PI/180))*@TAN($C$15*@PI/180))

D52: (F2) PR ($C$14*$G$24+($C$41*@COS(A52)-$E$40-$E$39*@SIN(A52)+$C$25*@COS
($C$26*@PI/180))*@TAN($C$15*@PI/180))

E52: (F2) PR ($C$14*$G$24+($C$41*@COS(A52)-$F$40-$F$39*@SIN(A52)+$C$25*@COS
($C$26*@PI/180))*@TAN($C$15*@PI/180))

G52: PR [W9] +A52/@PI*180

A53: (F3) PR (+C8+C10)/2

B53: (F2) PR ($C$14*$G$24+($C$41*@COS(A53)-$C$40-$C$39*@SIN(A53)+$C$25*@COS
($C$26*@PI/180))*@TAN($C$15*@PI/180))

C53: (F2) PR ($C$14*$G$24+($C$41*@COS(A53)-$D$40-$D$39*@SIN(A53)+$C$25*@COS
($C$26*@PI/180))*@TAN($C$15*@PI/180))

D53: (F2) PR ($C$14*$G$24+($C$41*@COS(A53)-$E$40-$E$39*@SIN(A53)+$C$25*@COS
($C$26*@PI/180))*@TAN($C$15*@PI/180))

E53: (F2) PR ($C$14*$G$24+($C$41*@COS(A53)-$F$40-$F$39*@SIN(A53)+$C$25*@COS
($C$26*@PI/180))*@TAN($C$15*@PI/180))

G53: PR [W9] +A53/@PI*180

A54: (F3) PR +C10

B54: (F2) PR ($C$14*$G$24+($C$41*@COS(A54)-$C$40-$C$39*@SIN(A54)+$C$25*@COS
($C$26*@PI/180))*@TAN($C$15*@PI/180))

C54: (F2) PR ($C$14*$G$24+($C$41*@COS(A54)-$D$40-$D$39*@SIN(A54)+$C$25*@COS
($C$26*@PI/180))*@TAN($C$15*@PI/180))

D54: (F2) PR ($C$14*$G$24+($C$41*@COS(A54)-$E$40-$E$39*@SIN(A54)+$C$25*@COS
($C$26*@PI/180))*@TAN($C$15*@PI/180))

E54: (F2) PR ($C$14*$G$24+($C$41*@COS(A54)-$F$40-$F$39*@SIN(A54)+$C$25*@COS
($C$26*@PI/180))*@TAN($C$15*@PI/180))

G54: PR [W9] +A54/@PI*180

A55: PR \+

B55: PR \+

C55: PR \+

D55: PR \+

E55: PR \+

F55: PR [W9] \+

G55: PR [W9] \+

A56: PR 'RADIAN
B56: PR 'G R O U N D W A T E R   C O N D I T I O N S
G56: PR [W9] 'DEGREE
A57: PR 'Plane
B57: PR "A
C57: PR "B
D57: PR "C
E57: PR "D
G57: PR [W9] 'Plane
A58: PR 'Angle
B58: PR 'A C T I V A T I N G   F O R C E S (kN/m)
G58: PR [W9] 'Angle
A59: PR \+
B59: PR \+
C59: PR \+
D59: PR \+
E59: PR \+
F59: PR [W9] \+
G59: PR [W9] \+
A60: (F3) PR +A50
B60: (F2) PR ($C$41*@SIN(A60)+$C$39*@COS(A60)-$C$25*@SIN($C$26*@PI/180))
C60: (F2) PR ($C$41*@SIN(A60)+$D$39*@COS(A60)-$C$25*@SIN($C$26*@PI/180))
D60: (F2) PR ($C$41*@SIN(A60)+$E$39*@COS(A60)-$C$25*@SIN($C$26*@PI/180))
E60: (F2) PR ($C$41*@SIN(A60)+$F$39*@COS(A60)-$C$25*@SIN($C$26*@PI/180))
G60: PR [W9] +A60/@PI*180
A61: (F3) PR +A51
B61: (F2) PR ($C$41*@SIN(A61)+$C$39*@COS(A61)-$C$25*@SIN($C$26*@PI/180))
C61: (F2) PR ($C$41*@SIN(A61)+$D$39*@COS(A61)-$C$25*@SIN($C$26*@PI/180))
D61: (F2) PR ($C$41*@SIN(A61)+$E$39*@COS(A61)-$C$25*@SIN($C$26*@PI/180))
E61: (F2) PR ($C$41*@SIN(A61)+$F$39*@COS(A61)-$C$25*@SIN($C$26*@PI/180))
G61: PR [W9] +A61/@PI*180
A62: (F3) PR +A52
B62: (F2) PR ($C$41*@SIN(A62)+$C$39*@COS(A62)-$C$25*@SIN($C$26*@PI/180))
C62: (F2) PR ($C$41*@SIN(A62)+$D$39*@COS(A62)-$C$25*@SIN($C$26*@PI/180))
D62: (F2) PR ($C$41*@SIN(A62)+$E$39*@COS(A62)-$C$25*@SIN($C$26*@PI/180))
E62: (F2) PR ($C$41*@SIN(A62)+$F$39*@COS(A62)-$C$25*@SIN($C$26*@PI/180))
G62: PR [W9] +A62/@PI*180
A63: (F3) PR +A53
B63: (F2) PR ($C$41*@SIN(A63)+$C$39*@COS(A63)-$C$25*@SIN($C$26*@PI/180))
C63: (F2) PR ($C$41*@SIN(A63)+$D$39*@COS(A63)-$C$25*@SIN($C$26*@PI/180))
D63: (F2) PR ($C$41*@SIN(A63)+$E$39*@COS(A63)-$C$25*@SIN($C$26*@PI/180))
E63: (F2) PR ($C$41*@SIN(A63)+$F$39*@COS(A63)-$C$25*@SIN($C$26*@PI/180))
G63: PR [W9] +A63/@PI*180
A64: (F3) PR +A54
B64: (F2) PR ($C$41*@SIN(A64)+$C$39*@COS(A64)-$C$25*@SIN($C$26*@PI/180))
C64: (F2) PR ($C$41*@SIN(A64)+$D$39*@COS(A64)-$C$25*@SIN($C$26*@PI/180))
D64: (F2) PR ($C$41*@SIN(A64)+$E$39*@COS(A64)-$C$25*@SIN($C$26*@PI/180))
E64: (F2) PR ($C$41*@SIN(A64)+$F$39*@COS(A64)-$C$25*@SIN($C$26*@PI/180))
G64: PR [W9] +A64/@PI*180
A67: PR \+
B67: PR \+
C67: PR \+
D67: PR \+
E67: PR \+
F67: PR [W9] \+
G67: PR [W9] \+
A68: PR 'RADIAN

B68: PR 'G R O U N D W A T E R   C O N D I T I O N S
G68: PR [W9] 'DEGREE
A69: PR 'Plane
B69: PR "A
C69: PR "B
D69: PR "C
E69: PR "D
G69: PR [W9] 'Plane
A70: PR 'Angle
C70: PR 'F A C T O R   O F   S A F E T Y
G70: PR [W9] 'Angle
A71: PR \+
B71: PR \+
C71: PR \+
D71: PR \+
E71: PR \+
F71: PR [W9] \+
G71: PR [W9] \+
A72: (F3) PR +A50
B72: (F2) PR +B50/B60
C72: (F2) PR +C50/C60
D72: (F2) PR +D50/D60
E72: (F2) PR +E50/E60
F72: (H) PR [W9] (@MAX($B$72..$E$76)-@MIN($B$72..$E$76))/5+@MAX($B$72..$E$76)
G72: PR [W9] +A72/@PI*180
A73: (F3) PR +A51
B73: (F2) PR +B51/B61
C73: (F2) PR +C51/C61
D73: (F2) PR +D51/D61
E73: (F2) PR +E51/E61
F73: (H) PR [W9] (@MAX($B$72..$E$76)-@MIN($B$72..$E$76))/5+@MAX($B$72..$E$76)
G73: PR [W9] +A73/@PI*180
A74: (F3) PR +A52
B74: (F2) PR +B52/B62
C74: (F2) PR +C52/C62
D74: (F2) PR +D52/D62
E74: (F2) PR +E52/E62
F74: (H) PR [W9] (@MAX($B$72..$E$76)-@MIN($B$72..$E$76))/5+@MAX($B$72..$E$76)
G74: PR [W9] +A74/@PI*180
A75: (F3) PR +A53
B75: (F2) PR +B53/B63
C75: (F2) PR +C53/C63
D75: (F2) PR +D53/D63
E75: (F2) PR +E53/E63
F75: (H) PR [W9] (@MAX($B$72..$E$76)-@MIN($B$72..$E$76))/5+@MAX($B$72..$E$76)
G75: PR [W9] +A75/@PI*180
A76: (F3) PR +A54
B76: (F2) PR +B54/B64
C76: (F2) PR +C54/C64
D76: (F2) PR +D54/D64
E76: (F2) PR +E54/E64
F76: (H) PR [W9] (@MAX($B$72..$E$76)-@MIN($B$72..$E$76))/5+@MAX($B$72..$E$76)
G76: PR [W9] +A76/@PI*180
A77: PR \+
B77: PR \+
C77: PR \+
D77: PR \+

E77: PR \+
F77: PR [W9] \+
G77: PR [W9] \+

7. FILE: PILECOH

D1: PR [W11] 'DESIGN OF PILES
D2: PR [W11] *
E2: PR [W13] *
A4: PR [W12] 'The program considers the following :
C5: PR [W11] 'pile type (bored or driven)
C6: PR [W11] 'soil type (cohesive or cohesionless)
C7: PR [W11] 'variation of pile capacity with length and size
A9: PR [W12] 'Design Considerations :
F9: PR [W9] 'Design Conditions
A10: PR [W12] \-
B10: PR [W13] \-
F10: PR [W9] \-
G10: PR [W10] \-
A11: PR [W12] 'Pile Type: Driven (H)igh or (L)ow displacement, or (B)ored ?
F11: U [W9] "b
D12: PR [W11] 'Pile size (shaft) - d
F12: U [W9] 1200
G12: PR [W10] 'mm
D13: PR [W11] 'Pile size (Base)
F13: U [W9] 1200
G13: PR [W10] 'mm
C14: PR [W11] 'Pile Shape - (R)ound or (S)quare ?
F14: U [W9] "r
D15: PR [W11] 'Estimated pile load
F15: U [W9] 1500
G15: PR [W10] 'KN
C17: PR [W11] 'Factor of safety (Bearing)
F17: U [W9] 3
C18: PR [W11] 'Factor of safety (Shaft)
F18: U [W9] 1.5
C19: PR [W11] 'Factor of safety (Overall)
F19: U [W9] 2.5
A21: PR [W12] 'Soil Type - (C)ohesive clays or (N)on-cohesive ?
F21: U [W9] "c
A22: PR [W12] @IF(F21="c","Is the degree of consolidation of the clays","Is the Non-cohesive soil")
F22: U [W9] "h
A23: PR [W12] @IF(F21="n","Gravel and (C)oarse Grained Sand or Non-plastic (S)ilt ?",
"(N)ormal, (L)ightly or (H)ighly Overconsolidated, or (F)issured ?")
C24: PR [W11] 'Strength paramaters at depth of
F24: U [W9] 1.5
G24: PR [W10] 'metres
E25: PR [W13] 'Increment
F25: U [W9] 1.5
G25: PR [W10] 'metres
A27: PR [W12] @IF(F11="b","Design of Bored piles in","Design of Driven piles in")
C27: PR [W11] @IF(F21="n",@IF(F22="c","coarse grained sand / gravel","non plastic silt"),@IF
(F22="n","normally consolidated clays",@IF(F22="L","lightly overconsolidated clays",@IF(F22="H",
"highly overconsolidated clays","fissured clays"))))

A28: PR [W12] \-
B28: PR [W13] \-
C28: PR [W11] \-
D28: PR [W11] \-
E28: PR [W13] \-
F28: PR [W9] \-
G28: PR [W10] \-
H28: PR [W10] \-
I28: PR \-
J28: PR \-
K28: PR \-
L28: PR \-
M28: PR 'Single
B29: PR [W13] @IF(F21="c","Undrained","Standard")
C29: PR [W11] "Embedment
D29: PR [W11] @IF(F21="c","Adhesion","Average")
E29: PR [W13] ^Max. Bearing
F29: PR [W9] ^Bearing
G29: PR [W10] 'Ult Bearing Capacity
I29: PR ' Ultimate Load
K29: PR 'Total
L29: PR ^No of
A30: PR [W12] ^Depth
B30: PR [W13] @IF(F21="c","Shear","Penetration")
C30: PR [W11] ^Ratio
D30: PR [W11] @IF(F21="c","Coefficient","Over")
E30: PR [W13] ^Capacity
F30: PR [W9] ^Capacity
G30: PR [W10] "(kN/m/m)
I30: PR "(kN)
K30: PR 'Allowable
L30: PR ^Piles
M30: PR ^Pile
A31: PR [W12] ^(metres)
B31: PR [W13] @IF(F21="c","Strength","Test")
D31: PR [W11] @IF(F21="c",@IF(F11="b","of Bored","of Driven"),"Length")
E31: PR [W13] ^Factor
F31: PR [W9] ^Factor
G31: PR [W10] ^Base
H31: PR [W10] ^Shaft
I31: PR ^Base
J31: PR ^Shaft
K31: PR 'Load (kN)
L31: PR ' required
A32: PR [W12] ^L
B32: PR [W13] @IF(F21="c","Cu (kPa)","(N value)")
C32: PR [W11] ^L/d
D32: PR [W11] @IF(F21="c","Piles","(N aveg)")
E32: PR [W13] @IF(F21="c","Nc (max)",@IF(F22="c","40 L /d","13 L/d"))
F32: PR [W9] ^used
G32: PR [W10] ^q(b)
H32: PR [W10] ^q(s)
I32: PR ^Q(b)
J32: PR ^Q(s)
K32: PR 'Q (all)
M32: PR ^Capacity
A33: PR [W12] \-

B33: PR [W13] \-
C33: PR [W11] \-
D33: PR [W11] \-
E33: PR [W13] \-
F33: PR [W9] \-
G33: PR [W10] \-
H33: PR [W10] \-
I33: PR \-
J33: PR \-
K33: PR \-
L33: PR \-
M33: PR \-
A34: PR [W12] +F24
B34: U [W13] 75
C34: (F1) PR [W11] +A34/($F$12/1000)
D34: (F2) PR [W11] @IF($F$21="c",@IF($F$11="b",@IF($F$22="f",0.3,0.45),@IF(B34<25,1,@IF(B34>100,0.3,0.5))),@AVG($B$34..B34))
E34: (F1) PR [W13] @IF($F$21="c",@IF($F$22="n",5,@IF($F$22="l",7,@IF($F$22="h",9,0.75*9))),@IF($F$11="b",13*C34,40*C34))
F34: (F1) PR [W9] @IF($F$21="c",@IF(C34>4,E34,2*@PI),@IF($F$11="b",@IF($F$22="c",@MIN(E34,130),@MIN(E34,100)),@IF($F$22="c",@MIN(E34,400),@MIN(E34,300))))
G34: (F1) PR [W10] +F34*B34
H34: (F1) PR [W10] @IF($F$21="c",B34*D34,@IF($F$11="h",2*D34,D34))
I34: (F1) PR @IF($F$14="r",G34*@PI*$F$13^2/1000^2/4,G34*$F$13^2/1000^2)
J34: (F1) PR @IF($F$14="r",+J33+H34*(A34-A33)*(@PI*$F$12/1000),+J33+H34*(A34-A33)*(4*$F$12/1000))
K34: (F1) PR @MIN(I34/$F$17+J34/$F$18,(I34+J34)/$F$19)
L34: PR @INT(K34/$F$15)+1
M34: PR +$F$15
A35: PR [W12] +A34+$F$25
B35: U [W13] 125
C35: (F1) PR [W11] +A35/($F$12/1000)
D35: (F2) PR [W11] @IF($F$21="c",@IF($F$11="b",@IF($F$22="f",0.3,0.45),@IF(B35<25,1,@IF(B35>100,0.3,0.5))),@AVG($B$34..B35))
E35: (F1) PR [W13] @IF($F$21="c",@IF($F$22="n",5,@IF($F$22="l",7,@IF($F$22="h",9,0.75*9))),@IF($F$11="b",13*C35,40*C35))
F35: (F1) PR [W9] @IF($F$21="c",@IF(C35>4,E35,2*@PI),@IF($F$11="b",@IF($F$22="c",@MIN(E35,130),@MIN(E35,100)),@IF($F$22="c",@MIN(E35,400),@MIN(E35,300))))
G35: (F1) PR [W10] +F35*B35
H35: (F1) PR [W10] @IF($F$21="c",B35*D35,@IF($F$11="h",2*D35,D35))
I35: (F1) PR @IF($F$14="r",G35*@PI*$F$13^2/1000^2/4,G35*$F$13^2/1000^2)
J35: (F1) PR @IF($F$14="r",+J34+H35*(A35-A34)*(@PI*$F$12/1000),+J34+H35*(A35-A34)*(4*$F$12/1000))
K35: (F1) PR @MIN(I35/$F$17+J35/$F$18,(I35+J35)/$F$19)
L35: PR @INT(K35/$F$15)+1
M35: PR +$F$15
A36: PR [W12] +A35+$F$25
B36: U [W13] 150
C36: (F1) PR [W11] +A36/($F$12/1000)
D36: (F2) PR [W11] @IF($F$21="c",@IF($F$11="b",@IF($F$22="f",0.3,0.45),@IF(B36<25,1,@IF(B36>100,0.3,0.5))),@AVG($B$34..B36))
E36: (F1) PR [W13] @IF($F$21="c",@IF($F$22="n",5,@IF($F$22="l",7,@IF($F$22="h",9,0.75*9))),@IF($F$11="b",13*C36,40*C36))
F36: (F1) PR [W9] @IF($F$21="c",@IF(C36>4,E36,2*@PI),@IF($F$11="b",@IF($F$22="c",@MIN(E36,130),@MIN(E36,100)),@IF($F$22="c",@MIN(E36,400),@MIN(E36,300))))
G36: (F1) PR [W10] +F36*B36
H36: (F1) PR [W10] @IF($F$21="c",B36*D36,@IF($F$11="h",2*D36,D36))

I36: (F1) PR @IF($F$14="r",G36*@PI*$F$13^2/1000^2/4,G36*$F$13^2/1000^2)

J36: (F1) PR @IF($F$14="r",+J35+H36*(A36-A35)*(@PI*$F$12/1000),+J35+H36*(A36-A35)*
(4*$F$12/1000))

K36: (F1) PR @MIN(I36/$F$17+J36/$F$18,(I36+J36)/$F$19)

L36: PR @INT(K36/$F$15)+1

M36: PR +$F$15

A37: PR [W12] +A36+$F$25

B37: U [W13] 100

C37: (F1) PR [W11] +A37/($F$12/1000)

D37: (F2) PR [W11] @IF($F$21="c",@IF($F$11="b",@IF($F$22="f",0.3,0.45),@IF(B37<25,1,@IF
(B37>100,0.3,0.5))),@AVG($B$34..B37))

E37: (F1) PR [W13] @IF($F$21="c",@IF($F$22="n",5,@IF($F$22="l",7,@IF($F$22="h",9,0.75*
9))),@IF($F$11="b",13*C37,40*C37))

F37: (F1) PR [W9] @IF($F$21="c",@IF(C37>4,E37,2*@PI),@IF($F$11="b",@IF($F$22="c",@
MIN(E37,130),@MIN(E37,100)),@IF($F$22="c",@MIN(E37,400),@MIN(E37,300))))

G37: (F1) PR [W10] +F37*B37

H37: (F1) PR [W10] @IF($F$21="c",B37*D37,@IF($F$11="h",2*D37,D37))

I37: (F1) PR @IF($F$14="r",G37*@PI*$F$13^2/1000^2/4,G37*$F$13^2/1000^2)

J37: (F1) PR @IF($F$14="r",+J36+H37*(A37-A36)*(@PI*$F$12/1000),+J36+H37*(A37-A36)*
(4*$F$12/1000))

K37: (F1) PR @MIN(I37/$F$17+J37/$F$18,(I37+J37)/$F$19)

L37: PR @INT(K37/$F$15)+1

M37: PR +$F$15

A38: PR [W12] +A37+$F$25

B38: U [W13] 125

C38: (F1) PR [W11] +A38/($F$12/1000)

D38: (F2) PR [W11] @IF($F$21="c",@IF($F$11="b",@IF($F$22="f",0.3,0.45),@IF(B38<25,1,@IF
(B38>100,0.3,0.5))),@AVG($B$34..B38))

E38: (F1) PR [W13] @IF($F$21="c",@IF($F$22="n",5,@IF($F$22="l",7,@IF($F$22="h",9,0.75*
9))),@IF($F$11="b",13*C38,40*C38))

F38: (F1) PR [W9] @IF($F$21="c",@IF(C38>4,E38,2*@PI),@IF($F$11="b",@IF($F$22="c",@
MIN(E38,130),@MIN(E38,100)),@IF($F$22="c",@MIN(E38,400),@MIN(E38,300))))

G38: (F1) PR [W10] +F38*B38

H38: (F1) PR [W10] @IF($F$21="c",B38*D38,@IF($F$11="h",2*D38,D38))

I38: (F1) PR @IF($F$14="r",G38*@PI*$F$13^2/1000^2/4,G38*$F$13^2/1000^2)

J38: (F1) PR @IF($F$14="r",+J37+H38*(A38-A37)*(@PI*$F$12/1000),+J37+H38*(A38-A37)*
(4*$F$12/1000))

K38: (F1) PR @MIN(I38/$F$17+J38/$F$18,(I38+J38)/$F$19)

L38: PR @INT(K38/$F$15)+1

M38: PR +$F$15

A39: PR [W12] +A38+$F$25

B39: U [W13] 150

C39: (F1) PR [W11] +A39/($F$12/1000)

D39: (F2) PR [W11] @IF($F$21="c",@IF($F$11="b",@IF($F$22="f",0.3,0.45),@IF(B39<25,1,@IF
(B39>100,0.3,0.5))),@AVG($B$34..B39))

E39: (F1) PR [W13] @IF($F$21="c",@IF($F$22="n",5,@IF($F$22="l",7,@IF($F$22="h",9,0.75*
9))),@IF($F$11="b",13*C39,40*C39))

F39: (F1) PR [W9] @IF($F$21="c",@IF(C39>4,E39,2*@PI),@IF($F$11="b",@IF($F$22="c",@
MIN(E39,130),@MIN(E39,100)),@IF($F$22="c",@MIN(E39,400),@MIN(E39,300))))

G39: (F1) PR [W10] +F39*B39

H39: (F1) PR [W10] @IF($F$21="c",B39*D39,@IF($F$11="h",2*D39,D39))

I39: (F1) PR @IF($F$14="r",G39*@PI*$F$13^2/1000^2/4,G39*$F$13^2/1000^2)

J39: (F1) PR @IF($F$14="r",+J38+H39*(A39-A38)*(@PI*$F$12/1000),+J38+H39*(A39-A38)*
(4*$F$12/1000))

K39: (F1) PR @MIN(I39/$F$17+J39/$F$18,(I39+J39)/$F$19)

L39: PR @INT(K39/$F$15)+1

M39: PR +$F$15

A40: PR [W12] +A39+$F$25

B40: U [W13] 200

C40: (F1) PR [W11] +A40/($F$12/1000)

D40: (F2) PR [W11] @IF($F$21="c",@IF($F$11="b",@IF($F$22="f",0.3,0.45),@IF(B40<25,1,@IF(B40>100,0.3,0.5))),@AVG($B$34..B40))

E40: (F1) PR [W13] @IF($F$21="c",@IF($F$22="n",5,@IF($F$22="l",7,@IF($F$22="h",9,0.75*9))),@IF($F$11="b",13*C40,40*C40))

F40: (F1) PR [W9] @IF($F$21="c",@IF(C40>4,E40,2*@PI),@IF($F$11="b",@IF($F$22="c",@MIN(E40,130),@MIN(E40,100)),@IF($F$22="c",@MIN(E40,400),@MIN(E40,300))))

G40: (F1) PR [W10] +F40*B40

H40: (F1) PR [W10] @IF($F$21="c",B40*D40,@IF($F$11="h",2*D40,D40))

I40: (F1) PR @IF($F$14="r",G40*@PI*$F$13^2/1000^2/4,G40*$F$13^2/1000^2)

J40: (F1) PR @IF($F$14="r",+J39+H40*(A40-A39)*(@PI*$F$12/1000),+J39+H40*(A40-A39)*(4*$F$12/1000))

K40: (F1) PR @MIN(I40/$F$17+J40/$F$18,(I40+J40)/$F$19)

L40: PR @INT(K40/$F$15)+1

M40: PR +$F$15

A41: PR [W12] +A40+$F$25

B41: U [W13] 250

C41: (F1) PR [W11] +A41/($F$12/1000)

D41: (F2) PR [W11] @IF($F$21="c",@IF($F$11="b",@IF($F$22="f",0.3,0.45),@IF(B41<25,1,@IF(B41>100,0.3,0.5))),@AVG($B$34..B41))

E41: (F1) PR [W13] @IF($F$21="c",@IF($F$22="n",5,@IF($F$22="l",7,@IF($F$22="h",9,0.75*9))),@IF($F$11="b",13*C41,40*C41))

F41: (F1) PR [W9] @IF($F$21="c",@IF(C41>4,E41,2*@PI),@IF($F$11="b",@IF($F$22="c",@MIN(E41,130),@MIN(E41,100)),@IF($F$22="c",@MIN(E41,400),@MIN(E41,300))))

G41: (F1) PR [W10] +F41*B41

H41: (F1) PR [W10] @IF($F$21="c",B41*D41,@IF($F$11="h",2*D41,D41))

I41: (F1) PR @IF($F$14="r",G41*@PI*$F$13^2/1000^2/4,G41*$F$13^2/1000^2)

J41: (F1) PR @IF($F$14="r",+J40+H41*(A41-A40)*(@PI*$F$12/1000),+J40+H41*(A41-A40)*(4*$F$12/1000))

K41: (F1) PR @MIN(I41/$F$17+J41/$F$18,(I41+J41)/$F$19)

L41: PR @INT(K41/$F$15)+1

M41: PR +$F$15

A42: PR [W12] +A41+$F$25

B42: U [W13] 250

C42: (F1) PR [W11] +A42/($F$12/1000)

D42: (F2) PR [W11] @IF($F$21="c",@IF($F$11="b",@IF($F$22="f",0.3,0.45),@IF(B42<25,1,@IF(B42>100,0.3,0.5))),@AVG($B$34..B42))

E42: (F1) PR [W13] @IF($F$21="c",@IF($F$22="n",5,@IF($F$22="l",7,@IF($F$22="h",9,0.75*9))),@IF($F$11="b",13*C42,40*C42))

F42: (F1) PR [W9] @IF($F$21="c",@IF(C42>4,E42,2*@PI),@IF($F$11="b",@IF($F$22="c",@MIN(E42,130),@MIN(E42,100)),@IF($F$22="c",@MIN(E42,400),@MIN(E42,300))))

G42: (F1) PR [W10] +F42*B42

H42: (F1) PR [W10] @IF($F$21="c",B42*D42,@IF($F$11="h",2*D42,D42))

I42: (F1) PR @IF($F$14="r",G42*@PI*$F$13^2/1000^2/4,G42*$F$13^2/1000^2)

J42: (F1) PR @IF($F$14="r",+J41+H42*(A42-A41)*(@PI*$F$12/1000),+J41+H42*(A42-A41)*(4*$F$12/1000))

K42: (F1) PR @MIN(I42/$F$17+J42/$F$18,(I42+J42)/$F$19)

L42: PR @INT(K42/$F$15)+1

M42: PR +$F$15

A43: PR [W12] +A42+$F$25

B43: U [W13] 300

C43: (F1) PR [W11] +A43/($F$12/1000)

D43: (F2) PR [W11] @IF($F$21="c",@IF($F$11="b",@IF($F$22="f",0.3,0.45),@IF(B43<25,1,@IF(B43>100,0.3,0.5))),@AVG($B$34..B43))

E43: (F1) PR [W13] @IF($F$21="c",@IF($F$22="n",5,@IF($F$22="l",7,@IF($F$22="h",9,0.75*

9))),@IF($F$11="b",13*C43,40*C43))

F43: (F1) PR [W9] @IF($F$21="c",@IF(C43>4,E43,2*@PI),@IF($F$11="b",@IF($F$22="c",@
MIN(E43,130),@MIN(E43,100)),@IF($F$22="c",@MIN(E43,400),@MIN(E43,300))))

G43: (F1) PR [W10] +F43*B43

H43: (F1) PR [W10] @IF($F$21="c",B43*D43,@IF($F$11="h",2*D43,D43))

I43: (F1) PR @IF($F$14="r",G43*@PI*$F$13^2/1000^2/4,G43*$F$13^2/1000^2)

J43: (F1) PR @IF($F$14="r",+J42+H43*(A43-A42)*(@PI*$F$12/1000),+J42+H43*(A43-A42)*
(4*$F$12/1000))

K43: (F1) PR @MIN(I43/$F$17+J43/$F$18,(I43+J43)/$F$19)

L43: PR @INT(K43/$F$15)+1

M43: PR +$F$15

A44: PR [W12] +A43+$F$25

B44: U [W13] 350

C44: (F1) PR [W11] +A44/($F$12/1000)

D44: (F2) PR [W11] @IF($F$21="c",@IF($F$11="b",@IF($F$22="f",0.3,0.45),@IF(B44<25,1,@IF
(B44>100,0.3,0.5))),@AVG($B$34..B44))

E44: (F1) PR [W13] @IF($F$21="c",@IF($F$22="n",5,@IF($F$22="l",7,@IF($F$22="h",9,0.75*
9))),@IF($F$11="b",13*C44,40*C44))

F44: (F1) PR [W9] @IF($F$21="c",@IF(C44>4,E44,2*@PI),@IF($F$11="b",@IF($F$22="c",@
MIN(E44,130),@MIN(E44,100)),@IF($F$22="c",@MIN(E44,400),@MIN(E44,300))))

G44: (F1) PR [W10] +F44*B44

H44: (F1) PR [W10] @IF($F$21="c",B44*D44,@IF($F$11="h",2*D44,D44))

I44: (F1) PR @IF($F$14="r",G44*@PI*$F$13^2/1000^2/4,G44*$F$13^2/1000^2)

J44: (F1) PR @IF($F$14="r",+J43+H44*(A44-A43)*(@PI*$F$12/1000),+J43+H44*(A44-A43)*
(4*$F$12/1000))

K44: (F1) PR @MIN(I44/$F$17+J44/$F$18,(I44+J44)/$F$19)

L44: PR @INT(K44/$F$15)+1

M44: PR +$F$15

A45: PR [W12] +A44+$F$25

B45: U [W13] 350

C45: (F1) PR [W11] +A45/($F$12/1000)

D45: (F2) PR [W11] @IF($F$21="c",@IF($F$11="b",@IF($F$22="f",0.3,0.45),@IF(B45<25,1,@IF
(B45>100,0.3,0.5))),@AVG($B$34..B45))

E45: (F1) PR [W13] @IF($F$21="c",@IF($F$22="n",5,@IF($F$22="l",7,@IF($F$22="h",9,0.75*
9))),@IF($F$11="b",13*C45,40*C45))

F45: (F1) PR [W9] @IF($F$21="c",@IF(C45>4,E45,2*@PI),@IF($F$11="b",@IF($F$22="c",@
MIN(E45,130),@MIN(E45,100)),@IF($F$22="c",@MIN(E45,400),@MIN(E45,300))))

G45: (F1) PR [W10] +F45*B45

H45: (F1) PR [W10] @IF($F$21="c",B45*D45,@IF($F$11="h",2*D45,D45))

I45: (F1) PR @IF($F$14="r",G45*@PI*$F$13^2/1000^2/4,G45*$F$13^2/1000^2)

J45: (F1) PR @IF($F$14="r",+J44+H45*(A45-A44)*(@PI*$F$12/1000),+J44+H45*(A45-A44)*
(4*$F$12/1000))

K45: (F1) PR @MIN(I45/$F$17+J45/$F$18,(I45+J45)/$F$19)

L45: PR @INT(K45/$F$15)+1

M45: PR +$F$15

A46: PR [W12] \-

B46: PR [W13] \-

C46: PR [W11] \-

D46: PR [W11] \-

E46: PR [W13] \-

F46: PR [W9] \-

G46: PR [W10] \-

H46: PR [W10] \-

I46: PR \-

J46: PR \-

K46: PR \-

L46: PR \-

M46: PR \-

8. FILE: PILEGRA

D1: PR [W11] 'DESIGN OF PILES
D2: PR [W11] *
E2: PR [W13] *
A4: PR [W12] 'The program considers the following :
C5: PR [W11] 'pile type (bored or driven)
C6: PR [W11] 'soil type (cohesive or cohesionless)
C7: PR [W11] 'variation of pile capacity with length and size
A9: PR [W12] 'Design Considerations :
F9: PR [W9] 'Design Conditions
A10: PR [W12] \-
B10: PR [W13] \-
F10: PR [W9] \-
G10: PR [W10] \-
A11: PR [W12] 'Pile Type: Driven (H)igh or (L)ow displacement, or (B)ored ?
F11: U [W9] "h
D12: PR [W11] 'Pile size (shaft) - d
F12: U [W9] 275
G12: PR [W10] 'mm
D13: PR [W11] 'Pile size (Base)
F13: U [W9] 275
G13: PR [W10] 'mm
C14: PR [W11] 'Pile Shape - (R)ound or (S)quare ?
F14: U [W9] "s
D15: PR [W11] 'Estimated pile load
F15: U [W9] 1000
G15: PR [W10] 'KN
C17: PR [W11] 'Factor of safety (Bearing)
F17: U [W9] 3
C18: PR [W11] 'Factor of safety (Shaft)
F18: U [W9] 1.5
C19: PR [W11] 'Factor of safety (Overall)
F19: U [W9] 2.5
A21: PR [W12] 'Soil Type - (C)ohesive clays or (N)on-cohesive ?
F21: U [W9] "n
A22: PR [W12] @IF(F21="c","Is the degree of consolidation of the clays","Is the Non-cohesive
soil")
F22: U [W9] "c
A23: PR [W12] @IF(F21="n","Gravel and (C)oarse Grained Sand or Non-plastic (S)ilt ?",
"(N)ormal, (L)ightly or (H)ighly Overconsolidated, or (F)issured ?")
C24: PR [W11] 'Strength paramaters at depth of
F24: U [W9] 1.5
G24: PR [W10] 'metres
E25: PR [W13] 'Increment
F25: U [W9] 1.5
G25: PR [W10] 'metres
A27: PR [W12] @IF(F11="b","Design of Bored piles in","Design of Driven piles in")
C27: PR [W11] @IF(F21="n",@IF(F22="c","coarse grained sand / gravel","non plastic silt"),@IF
(F22="n","normally consolidated clays",@IF(F22="L","lightly overconsolidated clays",@IF(F22="H",
"highly overconsolidated clays","fissured clays"))))
A28: PR [W12] \-
B28: PR [W13] \-

C28: PR [W11] \-
D28: PR [W11] \-
E28: PR [W13] \-
F28: PR [W9] \-
G28: PR [W10] \-
H28: PR [W10] \-
I28: PR \-
J28: PR \-
K28: PR \-
L28: PR \-
M28: PR 'Single
B29: PR [W13] @IF(F21="c","Undrained","Standard")
C29: PR [W11] "Embedment
D29: PR [W11] @IF(F21="c","Adhesion","Average")
E29: PR [W13] ^Max. Bearing
F29: PR [W9] ^Bearing
G29: PR [W10] 'Ult Bearing Capacity
I29: PR ' Ultimate Load
K29: PR 'Total
L29: PR ^No of
A30: PR [W12] ^Depth
B30: PR [W13] @IF(F21="c","Shear","Penetration")
C30: PR [W11] ^Ratio
D30: PR [W11] @IF(F21="c","Coefficient","Over")
E30: PR [W13] ^Capacity
F30: PR [W9] ^Capacity
G30: PR [W10] "(kN/m/m)
I30: PR "(kN)
K30: PR 'Allowable
L30: PR ^Piles
M30: PR ^Pile
A31: PR [W12] ^(metres)
B31: PR [W13] @IF(F21="c","Strength","Test")
D31: PR [W11] @IF(F21="c",@IF(F11="b","of Bored","of Driven"),"Length")
E31: PR [W13] ^Factor
F31: PR [W9] ^Factor
G31: PR [W10] ^Base
H31: PR [W10] ^Shaft
I31: PR ^Base
J31: PR ^Shaft
K31: PR 'Load (kN)
L31: PR ' required
A32: PR [W12] ^L
B32: PR [W13] @IF(F21="c","Cu (kPa)","(N value)")
C32: PR [W11] ^L/d
D32: PR [W11] @IF(F21="c","Piles","(N aveg)")
E32: PR [W13] @IF(F21="c","Nc (max)",@IF(F22="c","40 L /d","13 L/d"))
F32: PR [W9] ^used
G32: PR [W10] ^q(b)
H32: PR [W10] ^q(s)
I32: PR ^Q(b)
J32: PR ^Q(s)
K32: PR 'Q (all)
M32: PR ^Capacity
A33: PR [W12] \-
B33: PR [W13] \-
C33: PR [W11] \-

D33: PR [W11] \-
E33: PR [W13] \-
F33: PR [W9] \-
G33: PR [W10] \-
H33: PR [W10] \-
I33: PR \-
J33: PR \-
K33: PR \-
L33: PR \-
M33: PR \-
A34: PR [W12] +F24
B34: U [W13] 6
C34: (F1) PR [W11] +A34/($F$12/1000)
D34: (F2) PR [W11] @IF($F$21="c",@IF($F$11="b",@IF($F$22="f",0.3,0.45),@IF(B34<25,1,@IF
(B34>100,0.3,0.5))),@AVG($B$34..B34))
E34: (F1) PR [W13] @IF($F$21="c",@IF($F$22="n",5,@IF($F$22="l",7,@IF($F$22="h",9,0.75*
9))),@IF($F$11="b",13*C34,40*C34))
F34: (F1) PR [W9] @IF($F$21="c",@IF(C34>4,E34,2*@PI),@IF($F$11="b",@IF($F$22="c",@
MIN(E34,130),@MIN(E34,100)),@IF($F$22="c",@MIN(E34,400),@MIN(E34,300))))
G34: (F1) PR [W10] +F34*B34
H34: (F1) PR [W10] @IF($F$21="c",B34*D34,@IF($F$11="h",2*D34,D34))
I34: (F1) PR @IF($F$14="r",G34*@PI*$F$13^2/1000^2/4,G34*$F$13^2/1000^2)
J34: (F1) PR @IF($F$14="r",+J33+H34*(A34-A33)*(@PI*$F$12/1000),+J33+H34*(A34-A33)*
(4*$F$12/1000))
K34: (F1) PR @MIN(I34/$F$17+J34/$F$18,(I34+J34)/$F$19)
L34: PR @INT(K34/$F$15)+1
M34: PR +$F$15
A35: PR [W12] +A34+$F$25
B35: U [W13] 15
C35: (F1) PR [W11] +A35/($F$12/1000)
D35: (F2) PR [W11] @IF($F$21="c",@IF($F$11="b",@IF($F$22="f",0.3,0.45),@IF(B35<25,1,@IF
(B35>100,0.3,0.5))),@AVG($B$34..B35))
E35: (F1) PR [W13] @IF($F$21="c",@IF($F$22="n",5,@IF($F$22="l",7,@IF($F$22="h",9,0.75*
9))),@IF($F$11="b",13*C35,40*C35))
F35: (F1) PR [W9] @IF($F$21="c",@IF(C35>4,E35,2*@PI),@IF($F$11="b",@IF($F$22="c",@
MIN(E35,130),@MIN(E35,100)),@IF($F$22="c",@MIN(E35,400),@MIN(E35,300))))
G35: (F1) PR [W10] +F35*B35
H35: (F1) PR [W10] @IF($F$21="c",B35*D35,@IF($F$11="h",2*D35,D35))
I35: (F1) PR @IF($F$14="r",G35*@PI*$F$13^2/1000^2/4,G35*$F$13^2/1000^2)
J35: (F1) PR @IF($F$14="r",+J34+H35*(A35-A34)*(@PI*$F$12/1000),+J34+H35*(A35-A34)*
(4*$F$12/1000))
K35: (F1) PR @MIN(I35/$F$17+J35/$F$18,(I35+J35)/$F$19)
L35: PR @INT(K35/$F$15)+1
M35: PR +$F$15
A36: PR [W12] +A35+$F$25
B36: U [W13] 18
C36: (F1) PR [W11] +A36/($F$12/1000)
D36: (F2) PR [W11] @IF($F$21="c",@IF($F$11="b",@IF($F$22="f",0.3,0.45),@IF(B36<25,1,@IF
(B36>100,0.3,0.5))),@AVG($B$34..B36))
E36: (F1) PR [W13] @IF($F$21="c",@IF($F$22="n",5,@IF($F$22="l",7,@IF($F$22="h",9,0.75*
9))),@IF($F$11="b",13*C36,40*C36))
F36: (F1) PR [W9] @IF($F$21="c",@IF(C36>4,E36,2*@PI),@IF($F$11="b",@IF($F$22="c",@
MIN(E36,130),@MIN(E36,100)),@IF($F$22="c",@MIN(E36,400),@MIN(E36,300))))
G36: (F1) PR [W10] +F36*B36
H36: (F1) PR [W10] @IF($F$21="c",B36*D36,@IF($F$11="h",2*D36,D36))
I36: (F1) PR @IF($F$14="r",G36*@PI*$F$13^2/1000^2/4,G36*$F$13^2/1000^2)
J36: (F1) PR @IF($F$14="r",+J35+H36*(A36-A35)*(@PI*$F$12/1000),+J35+H36*(A36-A35)*

(4*$F$12/1000))
  K36: (F1) PR @MIN(I36/$F$17+J36/$F$18,(I36+J36)/$F$19)
  L36: PR @INT(K36/$F$15)+1
  M36: PR +$F$15
  A37: PR [W12] +A36+$F$25
  B37: U [W13] 12
  C37: (F1) PR [W11] +A37/($F$12/1000)
  D37: (F2) PR [W11] @IF($F$21="c",@IF($F$11="b",@IF($F$22="f",0.3,0.45),@IF(B37<25,1,@IF
(B37>100,0.3,0.5))),@AVG($B$34..B37))
  E37: (F1) PR [W13] @IF($F$21="c",@IF($F$22="n",5,@IF($F$22="l",7,@IF($F$22="h",9,0.75*
9))),@IF($F$11="b",13*C37,40*C37))
  F37: (F1) PR [W9] @IF($F$21="c",@IF(C37>4,E37,2*@PI),@IF($F$11="b",@IF($F$22="c",@
MIN(E37,130),@MIN(E37,100)),@IF($F$22="c",@MIN(E37,400),@MIN(E37,300))))
  G37: (F1) PR [W10] +F37*B37
  H37: (F1) PR [W10] @IF($F$21="c",B37*D37,@IF($F$11="h",2*D37,D37))
  I37: (F1) PR @IF($F$14="r",G37*@PI*$F$13^2/1000^2/4,G37*$F$13^2/1000^2)
  J37: (F1) PR @IF($F$14="r",+J36+H37*(A37-A36)*(@PI*$F$12/1000),+J36+H37*(A37-A36)*
(4*$F$12/1000))
  K37: (F1) PR @MIN(I37/$F$17+J37/$F$18,(I37+J37)/$F$19)
  L37: PR @INT(K37/$F$15)+1
  M37: PR +$F$15
  A38: PR [W12] +A37+$F$25
  B38: U [W13] 23
  C38: (F1) PR [W11] +A38/($F$12/1000)
  D38: (F2) PR [W11] @IF($F$21="c",@IF($F$11="b",@IF($F$22="f",0.3,0.45),@IF(B38<25,1,@IF
(B38>100,0.3,0.5))),@AVG($B$34..B38))
  E38: (F1) PR [W13] @IF($F$21="c",@IF($F$22="n",5,@IF($F$22="l",7,@IF($F$22="h",9,0.75*
9))),@IF($F$11="b",13*C38,40*C38))
  F38: (F1) PR [W9] @IF($F$21="c",@IF(C38>4,E38,2*@PI),@IF($F$11="b",@IF($F$22="c",@
MIN(E38,130),@MIN(E38,100)),@IF($F$22="c",@MIN(E38,400),@MIN(E38,300))))
  G38: (F1) PR [W10] +F38*B38
  H38: (F1) PR [W10] @IF($F$21="c",B38*D38,@IF($F$11="h",2*D38,D38))
  I38: (F1) PR @IF($F$14="r",G38*@PI*$F$13^2/1000^2/4,G38*$F$13^2/1000^2)
  J38: (F1) PR @IF($F$14="r",+J37+H38*(A38-A37)*(@PI*$F$12/1000),+J37+H38*(A38-A37)*
(4*$F$12/1000))
  K38: (F1) PR @MIN(I38/$F$17+J38/$F$18,(I38+J38)/$F$19)
  L38: PR @INT(K38/$F$15)+1
  M38: PR +$F$15
  A39: PR [W12] +A38+$F$25
  B39: U [W13] 27
  C39: (F1) PR [W11] +A39/($F$12/1000)
  D39: (F2) PR [W11] @IF($F$21="c",@IF($F$11="b",@IF($F$22="f",0.3,0.45),@IF(B39<25,1,@IF
(B39>100,0.3,0.5))),@AVG($B$34..B39))
  E39: (F1) PR [W13] @IF($F$21="c",@IF($F$22="n",5,@IF($F$22="l",7,@IF($F$22="h",9,0.75*
9))),@IF($F$11="b",13*C39,40*C39))
  F39: (F1) PR [W9] @IF($F$21="c",@IF(C39>4,E39,2*@PI),@IF($F$11="b",@IF($F$22="c",@
MIN(E39,130),@MIN(E39,100)),@IF($F$22="c",@MIN(E39,400),@MIN(E39,300))))
  G39: (F1) PR [W10] +F39*B39
  H39: (F1) PR [W10] @IF($F$21="c",B39*D39,@IF($F$11="h",2*D39,D39))
  I39: (F1) PR @IF($F$14="r",G39*@PI*$F$13^2/1000^2/4,G39*$F$13^2/1000^2)
  J39: (F1) PR @IF($F$14="r",+J38+H39*(A39-A38)*(@PI*$F$12/1000),+J38+H39*(A39-A38)*
(4*$F$12/1000))
  K39: (F1) PR @MIN(I39/$F$17+J39/$F$18,(I39+J39)/$F$19)
  L39: PR @INT(K39/$F$15)+1
  M39: PR +$F$15
  A40: PR [W12] +A39+$F$25
  B40: U [W13] 32

C40: (F1) PR [W11] +A40/($F$12/1000)

D40: (F2) PR [W11] @IF($F$21="c",@IF($F$11="b",@IF($F$22="f",0.3,0.45),@IF(B40<25,1,@IF(B40>100,0.3,0.5))),@AVG($B$34..B40))

E40: (F1) PR [W13] @IF($F$21="c",@IF($F$22="n",5,@IF($F$22="l",7,@IF($F$22="h",9,0.75*9))),@IF($F$11="b",13*C40,40*C40))

F40: (F1) PR [W9] @IF($F$21="c",@IF(C40>4,E40,2*@PI),@IF($F$11="b",@IF($F$22="c",@MIN(E40,130),@MIN(E40,100)),@IF($F$22="c",@MIN(E40,400),@MIN(E40,300))))

G40: (F1) PR [W10] +F40*B40

H40: (F1) PR [W10] @IF($F$21="c",B40*D40,@IF($F$11="h",2*D40,D40))

I40: (F1) PR @IF($F$14="r",G40*@PI*$F$13^2/1000^2/4,G40*$F$13^2/1000^2)

J40: (F1) PR @IF($F$14="r",+J39+H40*(A40-A39)*(@PI*$F$12/1000),+J39+H40*(A40-A39)*(4*$F$12/1000))

K40: (F1) PR @MIN(I40/$F$17+J40/$F$18,(I40+J40)/$F$19)

L40: PR @INT(K40/$F$15)+1

M40: PR +$F$15

A41: PR [W12] +A40+$F$25

B41: U [W13] 38

C41: (F1) PR [W11] +A41/($F$12/1000)

D41: (F2) PR [W11] @IF($F$21="c",@IF($F$11="b",@IF($F$22="f",0.3,0.45),@IF(B41<25,1,@IF(B41>100,0.3,0.5))),@AVG($B$34..B41))

E41: (F1) PR [W13] @IF($F$21="c",@IF($F$22="n",5,@IF($F$22="l",7,@IF($F$22="h",9,0.75*9))),@IF($F$11="b",13*C41,40*C41))

F41: (F1) PR [W9] @IF($F$21="c",@IF(C41>4,E41,2*@PI),@IF($F$11="b",@IF($F$22="c",@MIN(E41,130),@MIN(E41,100)),@IF($F$22="c",@MIN(E41,400),@MIN(E41,300))))

G41: (F1) PR [W10] +F41*B41

H41: (F1) PR [W10] @IF($F$21="c",B41*D41,@IF($F$11="h",2*D41,D41))

I41: (F1) PR @IF($F$14="r",G41*@PI*$F$13^2/1000^2/4,G41*$F$13^2/1000^2)

J41: (F1) PR @IF($F$14="r",+J40+H41*(A41-A40)*(@PI*$F$12/1000),+J40+H41*(A41-A40)*(4*$F$12/1000))

K41: (F1) PR @MIN(I41/$F$17+J41/$F$18,(I41+J41)/$F$19)

L41: PR @INT(K41/$F$15)+1

M41: PR +$F$15

A42: PR [W12] +A41+$F$25

B42: U [W13] 43

C42: (F1) PR [W11] +A42/($F$12/1000)

D42: (F2) PR [W11] @IF($F$21="c",@IF($F$11="b",@IF($F$22="f",0.3,0.45),@IF(B42<25,1,@IF(B42>100,0.3,0.5))),@AVG($B$34..B42))

E42: (F1) PR [W13] @IF($F$21="c",@IF($F$22="n",5,@IF($F$22="l",7,@IF($F$22="h",9,0.75*9))),@IF($F$11="b",13*C42,40*C42))

F42: (F1) PR [W9] @IF($F$21="c",@IF(C42>4,E42,2*@PI),@IF($F$11="b",@IF($F$22="c",@MIN(E42,130),@MIN(E42,100)),@IF($F$22="c",@MIN(E42,400),@MIN(E42,300))))

G42: (F1) PR [W10] +F42*B42

H42: (F1) PR [W10] @IF($F$21="c",B42*D42,@IF($F$11="h",2*D42,D42))

I42: (F1) PR @IF($F$14="r",G42*@PI*$F$13^2/1000^2/4,G42*$F$13^2/1000^2)

J42: (F1) PR @IF($F$14="r",+J41+H42*(A42-A41)*(@PI*$F$12/1000),+J41+H42*(A42-A41)*(4*$F$12/1000))

K42: (F1) PR @MIN(I42/$F$17+J42/$F$18,(I42+J42)/$F$19)

L42: PR @INT(K42/$F$15)+1

M42: PR +$F$15

A43: PR [W12] +A42+$F$25

B43: U [W13] 55

C43: (F1) PR [W11] +A43/($F$12/1000)

D43: (F2) PR [W11] @IF($F$21="c",@IF($F$11="b",@IF($F$22="f",0.3,0.45),@IF(B43<25,1,@IF(B43>100,0.3,0.5))),@AVG($B$34..B43))

E43: (F1) PR [W13] @IF($F$21="c",@IF($F$22="n",5,@IF($F$22="l",7,@IF($F$22="h",9,0.75*9))),@IF($F$11="b",13*C43,40*C43))

F43: (F1) PR [W9] @IF($F$21="c",@IF(C43>4,E43,2*@PI),@IF($F$11="b",@IF($F$22="c",@

MIN(E43,130),@MIN(E43,100)),@IF($F$22="c",@MIN(E43,400),@MIN(E43,300))))

   G43: (F1) PR [W10] +F43*B43

   H43: (F1) PR [W10] @IF($F$21="c",B43*D43,@IF($F$11="h",2*D43,D43))

   I43: (F1) PR @IF($F$14="r",G43*@PI*$F$13^2/1000^2/4,G43*$F$13^2/1000^2)

   J43: (F1) PR @IF($F$14="r",+J42+H43*(A43-A42)*(@PI*$F$12/1000),+J42+H43*(A43-A42)*
(4*$F$12/1000))

   K43: (F1) PR @MIN(I43/$F$17+J43/$F$18,(I43+J43)/$F$19)

   L43: PR @INT(K43/$F$15)+1

   M43: PR +$F$15

   A44: PR [W12] +A43+$F$25

   B44: U [W13] 56

   C44: (F1) PR [W11] +A44/($F$12/1000)

   D44: (F2) PR [W11] @IF($F$21="c",@IF($F$11="b",@IF($F$22="f",0.3,0.45),@IF(B44<25,1,@IF
(B44>100,0.3,0.5))),@AVG($B$34..B44))

   E44: (F1) PR [W13] @IF($F$21="c",@IF($F$22="n",5,@IF($F$22="l",7,@IF($F$22="h",9,0.75*
9))),@IF($F$11="b",13*C44,40*C44))

   F44: (F1) PR [W9] @IF($F$21="c",@IF(C44>4,E44,2*@PI),@IF($F$11="b",@IF($F$22="c",@
MIN(E44,130),@MIN(E44,100)),@IF($F$22="c",@MIN(E44,400),@MIN(E44,300))))

   G44: (F1) PR [W10] +F44*B44

   H44: (F1) PR [W10] @IF($F$21="c",B44*D44,@IF($F$11="h",2*D44,D44))

   I44: (F1) PR @IF($F$14="r",G44*@PI*$F$13^2/1000^2/4,G44*$F$13^2/1000^2)

   J44: (F1) PR @IF($F$14="r",+J43+H44*(A44-A43)*(@PI*$F$12/1000),+J43+H44*(A44-A43)*
(4*$F$12/1000))

   K44: (F1) PR @MIN(I44/$F$17+J44/$F$18,(I44+J44)/$F$19)

   L44: PR @INT(K44/$F$15)+1

   M44: PR +$F$15

   A45: PR [W12] +A44+$F$25

   B45: U [W13] 48

   C45: (F1) PR [W11] +A45/($F$12/1000)

   D45: (F2) PR [W11] @IF($F$21="c",@IF($F$11="b",@IF($F$22="f",0.3,0.45),@IF(B45<25,1,@IF
(B45>100,0.3,0.5))),@AVG($B$34..B45))

   E45: (F1) PR [W13] @IF($F$21="c",@IF($F$22="n",5,@IF($F$22="l",7,@IF($F$22="h",9,0.75*
9))),@IF($F$11="b",13*C45,40*C45))

   F45: (F1) PR [W9] @IF($F$21="c",@IF(C45>4,E45,2*@PI),@IF($F$11="b",@IF($F$22="c",@
MIN(E45,130),@MIN(E45,100)),@IF($F$22="c",@MIN(E45,400),@MIN(E45,300))))

   G45: (F1) PR [W10] +F45*B45

   H45: (F1) PR [W10] @IF($F$21="c",B45*D45,@IF($F$11="h",2*D45,D45))

   I45: (F1) PR @IF($F$14="r",G45*@PI*$F$13^2/1000^2/4,G45*$F$13^2/1000^2)

   J45: (F1) PR @IF($F$14="r",+J44+H45*(A45-A44)*(@PI*$F$12/1000),+J44+H45*(A45-A44)*
(4*$F$12/1000))

   K45: (F1) PR @MIN(I45/$F$17+J45/$F$18,(I45+J45)/$F$19)

   L45: PR @INT(K45/$F$15)+1

   M45: PR +$F$15

   A46: PR [W12] \-

   B46: PR [W13] \-

   C46: PR [W11] \-

   D46: PR [W11] \-

   E46: PR [W13] \-

   F46: PR [W9] \-

   G46: PR [W10] \-

   H46: PR [W10] \-

   I46: PR \-

   J46: PR \-

   K46: PR \-

   L46: PR \-

   M46: PR \-

9. FILE: 2LPILE1

A14: PR [W12] 'Design Considerations :
F14: PR [W9] 'Design Conditions
A15: PR [W12] \-
B15: PR [W13] \-
F15: PR [W9] \-
G15: PR [W10] \-
A17: PR [W12] 'What type of pile is proposed - (D)riven or (B)ored ?
F17: U [W9] "b
D18: PR [W11] 'Pile size (shaft)
F18: U [W9] 950
G18: PR [W10] 'mm
D19: PR [W11] 'Pile size (Base)
F19: U [W9] 950
G19: PR [W10] 'mm
B20: PR [W13] 'Pile Shape: (R)ound, (S)quare or (O)ctogonal ?
F20: U [W9] "o
D21: PR [W11] 'Estimated pile load
F21: U [W9] 800
G21: PR [W10] 'kN
C22: PR [W12] "   Design Structural Capacity
F22: U [W9] 1000
G22: PR [W10] 'kN
C23: PR [W12] "   Reduced Level at Surface
F23: U [W9] 18.42
G23: PR [W10] 'metres
C25: PR [W12] 'Factor of safety (Bearing)
F25: U [W9] 3
C26: PR [W12] 'Factor of safety (Shaft)
F26: U [W9] 1.5
C27: PR [W12] 'Factor of safety (Overall)
F27: U [W9] 2.5
A29: PR [W12] 'Soil Type - (C)ohesive clays or (N)on-cohesive ?
F29: U [W9] "c
A30: PR [W12] @IF(F29="c","Is the degree of consolidation of the clays","Is the Non-cohesive
soil")
F30: U [W9] "f
A31: PR [W12] @IF(F29="n","Gravel and (C)oarse Grained Sand or Non-plastic (S)ilt ?",
"(N)ormal, (L)ightly or (H)ighly Overconsolidated, or (F)issured ?")
A32: PR [W12] \-
A33: PR [W12] @IF($F$17="d"#AND#$F$29="c"#AND#$F$30<>"n","Check Geotechnical Model
to obtain relevant adhesion factors","Geotechnical Model is considered one layered - Use Model 4")
A34: PR [W12] \-
B34: PR [W13] \-
C34: PR [W12] \-
D34: PR [W11] \-
E34: PR [W11] \-
F34: PR [W9] \-
G34: PR [W10] \-
A35: PR [W12] "Geotechnical Model
C35: PR [W12] 'l
D35: PR [W11] "Model 1 :
E35: PR [W11] 'Sand/gravel over Stiff Clay
A36: U [W12] 1
C36: PR [W12] 'l
D36: PR [W11] "Model 2 :

E36: PR [W11] 'Soft Clay over Stiff Clay
A37: PR [W12] @IF(A36=1,"Sand/Gravel thickness",@IF(A36=2,"Soft Clay thickness ","Upper layer thickness = 0"))
C37: PR [W12] '|
D37: PR [W11] "Model 3 :
E37: PR [W11] 'Only Stiff Clay
A38: U [W12] 1.8
B38: PR [W13] "metres
C38: PR [W12] '|
D38: PR [W11] "Model 4 :
E38: PR [W11] 'One Layered Soil Strata
A39: PR [W12] \-
B39: PR [W13] \-
C39: PR [W12] \-
D39: PR [W11] \-
E39: PR [W11] \-
F39: PR [W9] \-
G39: PR [W10] \-
A41: PR [W12] @IF(F17="d","Design of Driven piles in","Design of Bored piles in")
C41: PR [W12] @IF(F29="n",@IF(F30="c"," coarse grained sand / gravel"," non plastic silt"),@IF (F30="n"," normally consolidated clays",@IF(F30="L"," lightly overconsolidated clays",@IF(F30="H"," highly overconsolidated clays"," fissured clays"))))
A42: PR [W12] \-
B42: PR [W13] \-
C42: PR [W12] \-
D42: PR [W11] \-
E42: PR [W11] \-
F42: PR [W9] \-
G42: PR [W10] \-
H42: PR [W9] \-
I42: PR \-
J42: PR [W11] \-
K42: PR \-
L42: PR \-
M42: PR \-
N42: PR \-
C43: PR [W12] @IF(F29="c","Undrained","Standard")
D43: PR [W11] @IF(F29="c","Adhesion","Average")
E43: PR [W11] 'Ult Bearing capacity
G43: PR [W10] ' Ultimate Load
I43: PR 'Total
P43: PR "Check on embedment
A44: PR [W12] ^Depth
B44: PR [W13] 'Reduced
C44: PR [W12] @IF(F29="c","Shear","Penetration")
D44: PR [W11] @IF(F29="c","Coefficient","Over")
E44: PR [W11] "(kN/m/m)
G44: PR [W10] "(kN)
I44: PR 'Allowable
J44: PR [W11] ^No of
K44: PR ^Pile
L44: PR 'Percentage Carried by
P44: PR ^L/D
R44: PR 'Ult Brg Cap
A45: PR [W12] ^(metres)
B45: PR [W13] 'Level
C45: PR [W12] @IF(F29="c","Strength","Test")

D45: PR [W11] @IF(F29="c",@IF(F17="d","of Driven","of Bored"),"Length")
E45: PR [W11] ^Base
F45: PR [W9] ^Shaft
G45: PR [W10] ^Base
H45: PR [W9] ^Shaft
I45: PR 'Load (kN)
J45: PR [W11] ^Piles
L45: PR "Base
M45: PR "Shaft
N45: PR "Total
P45: PR "for that
R45: PR ^Base
B46: PR [W13] '(metres)
C46: PR [W12] @IF(F29="c","Cu (kPa)","(N value)")
D46: PR [W11] @IF(F29="c","Piles","(N aveg)")
E46: PR [W11] ^q(b)
F46: PR [W9] ^q(s)
G46: PR [W10] ^Q(b)
H46: PR [W9] ^Q(s)
I46: PR 'Q (all)
J46: PR [W11] ^required
K46: PR ^Capacity
N46: PR "Pile
P46: PR ^layer
Q46: PR "Nc
R46: PR ^q(b)
A47: PR [W12] \-
B47: PR [W13] \-
C47: PR [W12] \-
D47: PR [W11] \-
E47: PR [W11] \-
F47: PR [W9] \-
G47: PR [W10] \-
H47: PR [W9] \-
I47: PR \-
J47: PR [W11] \-
K47: PR \-
L47: PR \-
M47: PR \-
N47: PR \-
P47: PR \-
Q47: PR \-
R47: PR \-
A48: PR [W12] 0
B48: PR [W13] +F23
C48: (F2) PR [W12] 0.00001
A49: U [W12] 1.8
B49: PR [W13] +$B$48-A49
C49: U [W12] 0
D49: (F2) PR [W11] @IF($F$29="c",@IF($F$17="d",+I90,@IF($F$30="h",0.45,0.3)),@AVG
($C$49..C49))
E49: (F1) PR [W11] @IF($F$17="d",@IF($F$29="n",@IF($F$30="c",400*C49,300*C49),@IF(F
30="n",5*C49,@IF($F$30="L",6*C49,9*C49))),@IF($F$29="n",@IF($F$30="c",130*C49,100*C49),@
IF($F$30="n",5*C49,@IF($F$30="h",9*C49,6*C49))))
F49: (F1) PR [W9] @IF($F$29="c",C49*D49,@IF($F$17="b",D49,2*D49))
G49: (F1) PR [W10] @IF($F$20="r"#OR#$F$20="o",@MIN(E49,R49)*@PI*$F$19^2/1000^2/4,@
MIN(E49,R49)*$F$19^2/1000^2)

H49: (F1) PR [W9] @IF($F$20="r"#OR#$F$20="o",F49*(A49-A48)*@PI*$F$18/1000+H48,F49*
(A49-A48)*$F$18/1000+H48)

I49: (F1) PR @MIN(G49/$F$25+H49/$F$26,(G49+H49)/$F$27)

J49: PR [W11] @INT(I49/$F$21)+1

K49: PR +$F$21

L49: (P1) PR +G49/(G49+H49)

M49: (P1) PR +H49/(G49+H49)

N49: (P1) PR +I49/K49

P49: (F1) PR (A49-A48)/($F$18/1000)

Q49: (F1) PR @IF(C49=C48,9,@VLOOKUP(P49,$P$64..$Q$68,1))

R49: PR +Q49*C49

A50: U [W12] 4.42

B50: PR [W13] +$B$48-A50

C50: U [W12] 100

D50: (F2) PR [W11] @IF($F$29="c",@IF($F$17="d",+I91,@IF($F$30="h",0.45,0.3)),@AVG
($C$49..C50))

E50: (F1) PR [W11] @IF($F$17="d",@IF($F$29="n",@IF($F$30="c",400*C50,300*C50),@IF(F
30="n",5*C50,@IF($F$30="L",6*C50,9*C50))),@IF($F$29="n",@IF($F$30="c",130*C50,100*C50),@
IF($F$30="n",5*C50,@IF($F$30="h",9*C50,6*C50)))))

F50: (F1) PR [W9] @IF($F$29="c",C50*D50,@IF($F$17="b",D50,2*D50))

G50: (F1) PR [W10] @IF($F$20="r"#OR#$F$20="o",@MIN(E50,R50)*@PI*$F$19^2/1000^2/4,@
MIN(E50,R50)*$F$19^2/1000^2)

H50: (F1) PR [W9] @IF($F$20="r"#OR#$F$20="o",F50*(A50-A49)*@PI*$F$18/1000+H49,F50*
(A50-A49)*$F$18/1000+H49)

I50: (F1) PR @MIN(G50/$F$25+H50/$F$26,(G50+H50)/$F$27)

J50: PR [W11] @INT(I50/$F$21)+1

K50: PR +$F$21

L50: (P1) PR +G50/(G50+H50)

M50: (P1) PR +H50/(G50+H50)

N50: (P1) PR +I50/K50

P50: (F1) PR (A50-A49)/($F$18/1000)

Q50: (F1) PR @IF(C50=C49,9,@VLOOKUP(P50,$P$64..$Q$68,1))

R50: PR +Q50*C50

A51: U [W12] 8

B51: PR [W13] +$B$48-A51

C51: U [W12] 200

D51: (F2) PR [W11] @IF($F$29="c",@IF($F$17="d",+I92,@IF($F$30="h",0.45,0.3)),@AVG
($C$49..C51))

E51: (F1) PR [W11] @IF($F$17="d",@IF($F$29="n",@IF($F$30="c",400*C51,300*C51),@IF(F
30="n",5*C51,@IF($F$30="L",6*C51,9*C51))),@IF($F$29="n",@IF($F$30="c",130*C51,100*C51),@
IF($F$30="n",5*C51,@IF($F$30="h",9*C51,6*C51)))))

F51: (F1) PR [W9] @IF($F$29="c",C51*D51,@IF($F$17="b",D51,2*D51))

G51: (F1) PR [W10] @IF($F$20="r"#OR#$F$20="o",@MIN(E51,R51)*@PI*$F$19^2/1000^2/4,@
MIN(E51,R51)*$F$19^2/1000^2)

H51: (F1) PR [W9] @IF($F$20="r"#OR#$F$20="o",F51*(A51-A50)*@PI*$F$18/1000+H50,F51*
(A51-A50)*$F$18/1000+H50)

I51: (F1) PR @MIN(G51/$F$25+H51/$F$26,(G51+H51)/$F$27)

J51: PR [W11] @INT(I51/$F$21)+1

K51: PR +$F$21

L51: (P1) PR +G51/(G51+H51)

M51: (P1) PR +H51/(G51+H51)

N51: (P1) PR +I51/K51

P51: (F1) PR (A51-A50)/($F$18/1000)

Q51: (F1) PR @IF(C51=C50,9,@VLOOKUP(P51,$P$64..$Q$68,1))

R51: PR +Q51*C51

A52: U [W12] 8

B52: PR [W13] +$B$48-A52

C52: U [W12] 450

D52: (F2) PR [W11] @IF($F$29="c",@IF($F$17="d",+I93,@IF($F$30="h",0.45,0.3)),@AVG ($C$49..C52))

E52: (F1) PR [W11] @IF($F$17="d",@IF($F$29="n",@IF($F$30="c",400*C52,300*C52),@IF($F$30="n",5*C52,@IF($F$30="L",6*C52,9*C52))),@IF($F$29="n",@IF($F$30="c",130*C52,100*C52),@IF($F$30="n",5*C52,@IF($F$30="h",9*C52,6*C52))))

F52: (F1) PR [W9] @IF($F$29="c",C52*D52,@IF($F$17="b",D52,2*D52))

G52: (F1) PR [W10] @IF($F$20="r"#OR#$F$20="o",@MIN(E52,R52)*@PI*$F$19^2/1000^2/4,@MIN(E52,R52)*$F$19^2/1000^2)

H52: (F1) PR [W9] @IF($F$20="r"#OR#$F$20="o",F52*(A52-A51)*@PI*$F$18/1000+H51,F52*(A52-A51)*$F$18/1000+H51)

I52: (F1) PR @MIN(G52/$F$25+H52/$F$26,(G52+H52)/$F$27)

J52: PR [W11] @INT(I52/$F$21)+1

K52: PR +$F$21

L52: (P1) PR +G52/(G52+H52)

M52: (P1) PR +H52/(G52+H52)

N52: (P1) PR +I52/K52

P52: (F1) PR (A52-A51)/($F$18/1000)

Q52: (F1) PR @IF(C52=C51,9,@VLOOKUP(P52,$P$64..$Q$68,1))

R52: PR +Q52*C52

A53: U [W12] 9

B53: PR [W13] +$B$48-A53

C53: U [W12] 450

D53: (F2) PR [W11] @IF($F$29="c",@IF($F$17="d",+I94,@IF($F$30="h",0.45,0.3)),@AVG ($C$49..C53))

E53: (F1) PR [W11] @IF($F$17="d",@IF($F$29="n",@IF($F$30="c",400*C53,300*C53),@IF($F$30="n",5*C53,@IF($F$30="L",6*C53,9*C53))),@IF($F$29="n",@IF($F$30="c",130*C53,100*C53),@IF($F$30="n",5*C53,@IF($F$30="h",9*C53,6*C53))))

F53: (F1) PR [W9] @IF($F$29="c",C53*D53,@IF($F$17="b",D53,2*D53))

G53: (F1) PR [W10] @IF($F$20="r"#OR#$F$20="o",@MIN(E53,R53)*@PI*$F$19^2/1000^2/4,@MIN(E53,R53)*$F$19^2/1000^2)

H53: (F1) PR [W9] @IF($F$20="r"#OR#$F$20="o",F53*(A53-A52)*@PI*$F$18/1000+H52,F53*(A53-A52)*$F$18/1000+H52)

I53: (F1) PR @MIN(G53/$F$25+H53/$F$26,(G53+H53)/$F$27)

J53: PR [W11] @INT(I53/$F$21)+1

K53: PR +$F$21

L53: (P1) PR +G53/(G53+H53)

M53: (P1) PR +H53/(G53+H53)

N53: (P1) PR +I53/K53

P53: (F1) PR (A53-A52)/($F$18/1000)

Q53: (F1) PR @IF(C53=C52,9,@VLOOKUP(P53,$P$64..$Q$68,1))

R53: PR +Q53*C53

A54: U [W12] 9.5

B54: PR [W13] +$B$48-A54

C54: U [W12] 450

D54: (F2) PR [W11] @IF($F$29="c",@IF($F$17="d",+I95,@IF($F$30="h",0.45,0.3)),@AVG ($C$49..C54))

E54: (F1) PR [W11] @IF($F$17="d",@IF($F$29="n",@IF($F$30="c",400*C54,300*C54),@IF($F$30="n",5*C54,@IF($F$30="L",6*C54,9*C54))),@IF($F$29="n",@IF($F$30="c",130*C54,100*C54),@IF($F$30="n",5*C54,@IF($F$30="h",9*C54,6*C54))))

F54: (F1) PR [W9] @IF($F$29="c",C54*D54,@IF($F$17="b",D54,2*D54))

G54: (F1) PR [W10] @IF($F$20="r"#OR#$F$20="o",@MIN(E54,R54)*@PI*$F$19^2/1000^2/4,@MIN(E54,R54)*$F$19^2/1000^2)

H54: (F1) PR [W9] @IF($F$20="r"#OR#$F$20="o",F54*(A54-A53)*@PI*$F$18/1000+H53,F54*(A54-A53)*$F$18/1000+H53)

I54: (F1) PR @MIN(G54/$F$25+H54/$F$26,(G54+H54)/$F$27)

J54: PR [W11] @INT(I54/$F$21)+1

```
K54: PR +$F$21
L54: (P1) PR +G54/(G54+H54)
M54: (P1) PR +H54/(G54+H54)
N54: (P1) PR +I54/K54
P54: (F1) PR (A54-A53)/($F$18/1000)
Q54: (F1) PR @IF(C54=C53,9,@VLOOKUP(P54,$P$64..$Q$68,1))
R54: PR +Q54*C54
A55: U [W12] 10
B55: PR [W13] +$B$48-A55
C55: U [W12] 450
D55: (F2) PR [W11] @IF($F$29="c",@IF($F$17="d",+I96,@IF($F$30="h",0.45,0.3)),@AVG
($C$49..C55))
E55: (F1) PR [W11] @IF($F$17="d",@IF($F$29="n",@IF($F$30="c",400*C55,300*C55),@IF($F$
30="n",5*C55,@IF($F$30="L",6*C55,9*C55))),@IF($F$29="n",@IF($F$30="c",130*C55,100*C55),@
IF($F$30="n",5*C55,@IF($F$30="h",9*C55,6*C55))))
F55: (F1) PR [W9] @IF($F$29="c",C55*D55,@IF($F$17="b",D55,2*D55))
G55: (F1) PR [W10] @IF($F$20="r"#OR#$F$20="o",@MIN(E55,R55)*@PI*$F$19^2/1000^2/4,@
MIN(E55,R55)*$F$19^2/1000^2)
H55: (F1) PR [W9] @IF($F$20="r"#OR#$F$20="o",F55*(A55-A54)*@PI*$F$18/1000+H54,F55*
(A55-A54)*$F$18/1000+H54)
I55: (F1) PR @MIN(G55/$F$25+H55/$F$26,(G55+H55)/$F$27)
J55: PR [W11] @INT(I55/$F$21)+1
K55: PR +$F$21
L55: (P1) PR +G55/(G55+H55)
M55: (P1) PR +H55/(G55+H55)
N55: (P1) PR +I55/K55
P55: (F1) PR (A55-A54)/($F$18/1000)
Q55: (F1) PR @IF(C55=C54,9,@VLOOKUP(P55,$P$64..$Q$68,1))
R55: PR +Q55*C55
A56: U [W12] 10.5
B56: PR [W13] +$B$48-A56
C56: U [W12] 450
D56: (F2) PR [W11] @IF($F$29="c",@IF($F$17="d",+I97,@IF($F$30="h",0.45,0.3)),@AVG
($C$49..C56))
E56: (F1) PR [W11] @IF($F$17="d",@IF($F$29="n",@IF($F$30="c",400*C56,300*C56),@IF($F$
30="n",5*C56,@IF($F$30="L",6*C56,9*C56))),@IF($F$29="n",@IF($F$30="c",130*C56,100*C56),@
IF($F$30="n",5*C56,@IF($F$30="h",9*C56,6*C56))))
F56: (F1) PR [W9] @IF($F$29="c",C56*D56,@IF($F$17="b",D56,2*D56))
G56: (F1) PR [W10] @IF($F$20="r"#OR#$F$20="o",@MIN(E56,R56)*@PI*$F$19^2/1000^2/4,@
MIN(E56,R56)*$F$19^2/1000^2)
H56: (F1) PR [W9] @IF($F$20="r"#OR#$F$20="o",F56*(A56-A55)*@PI*$F$18/1000+H55,F56*
(A56-A55)*$F$18/1000+H55)
I56: (F1) PR @MIN(G56/$F$25+H56/$F$26,(G56+H56)/$F$27)
J56: PR [W11] @INT(I56/$F$21)+1
K56: PR +$F$21
L56: (P1) PR +G56/(G56+H56)
M56: (P1) PR +H56/(G56+H56)
N56: (P1) PR +I56/K56
P56: (F1) PR (A56-A55)/($F$18/1000)
Q56: (F1) PR @IF(C56=C55,9,@VLOOKUP(P56,$P$64..$Q$68,1))
R56: PR +Q56*C56
A57: U [W12] 11
B57: PR [W13] +$B$48-A57
C57: U [W12] 450
D57: (F2) PR [W11] @IF($F$29="c",@IF($F$17="d",+I98,@IF($F$30="h",0.45,0.3)),@AVG
($C$49..C57))
E57: (F1) PR [W11] @IF($F$17="d",@IF($F$29="n",@IF($F$30="c",400*C57,300*C57),@IF($F$
```

30="n",5*C57,@IF($F$30="L",6*C57,9*C57))),@IF($F$29="n",@IF($F$30="c",130*C57,100*C57),@
IF($F$30="n",5*C57,@IF($F$30="h",9*C57,6*C57))))

F57: (F1) PR [W9] @IF($F$29="c",C57*D57,@IF($F$17="b",D57,2*D57))

G57: (F1) PR [W10] @IF($F$20="r"#OR#$F$20="o",@MIN(E57,R57)*@PI*$F$19^2/1000^2/4,@
MIN(E57,R57)*$F$19^2/1000^2)

H57: (F1) PR [W9] @IF($F$20="r"#OR#$F$20="o",F57*(A57-A56)*@PI*$F$18/1000+H56,F57*
(A57-A56)*$F$18/1000+H56)

I57: (F1) PR @MIN(G57/$F$25+H57/$F$26,(G57+H57)/$F$27)

J57: PR [W11] @INT(I57/$F$21)+1

K57: PR +$F$21

L57: (P1) PR +G57/(G57+H57)

M57: (P1) PR +H57/(G57+H57)

N57: (P1) PR +I57/K57

P57: (F1) PR (A57-A56)/($F$18/1000)

Q57: (F1) PR @IF(C57=C56,9,@VLOOKUP(P57,$P$64..$Q$68,1))

R57: PR +Q57*C57

A58: U [W12] 11.5

B58: PR [W13] +$B$48-A58

C58: U [W12] 450

D58: (F2) PR [W11] @IF($F$29="c",@IF($F$17="d",+I99,@IF($F$30="h",0.45,0.3)),@AVG
($C$49..C58))

E58: (F1) PR [W11] @IF($F$17="d",@IF($F$29="n",@IF($F$30="c",400*C58,300*C58),@IF(F
30="n",5*C58,@IF($F$30="L",6*C58,9*C58))),@IF($F$29="n",@IF($F$30="c",130*C58,100*C58),@
IF($F$30="n",5*C58,@IF($F$30="h",9*C58,6*C58))))

F58: (F1) PR [W9] @IF($F$29="c",C58*D58,@IF($F$17="b",D58,2*D58))

G58: (F1) PR [W10] @IF($F$20="r"#OR#$F$20="o",@MIN(E58,R58)*@PI*$F$19^2/1000^2/4,@
MIN(E58,R58)*$F$19^2/1000^2)

H58: (F1) PR [W9] @IF($F$20="r"#OR#$F$20="o",F58*(A58-A57)*@PI*$F$18/1000+H57,F58*
(A58-A57)*$F$18/1000+H57)

I58: (F1) PR @MIN(G58/$F$25+H58/$F$26,(G58+H58)/$F$27)

J58: PR [W11] @INT(I58/$F$21)+1

K58: PR +$F$21

L58: (P1) PR +G58/(G58+H58)

M58: (P1) PR +H58/(G58+H58)

N58: (P1) PR +I58/K58

P58: (F1) PR (A58-A57)/($F$18/1000)

Q58: (F1) PR @IF(C58=C57,9,@VLOOKUP(P58,$P$64..$Q$68,1))

R58: PR +Q58*C58

A59: PR [W12] \-

B59: PR [W13] \-

C59: PR [W12] \-

D59: PR [W11] \-

E59: PR [W11] \-

F59: PR [W9] \-

G59: PR [W10] \-

H59: PR [W9] \-

I59: PR \-

J59: PR [W11] \-

K59: PR \-

L59: PR \-

M59: PR \-

N59: PR \-

P59: PR \-

Q59: PR \-

R59: PR \-

A60: U [W12] 'Borehole 2

Q61: PR "Brg Capacity factor

A62: PR [W12] @IF($F$17="d"#AND#$F$29="c"#AND#$F$30<>"n","Check Geotechnical Model
to obtain relevant adhesion factors","Geotechnical Model is considered one layered")
P62: PR "L/D
Q62: PR "Nc
A63: PR [W12] \+
B63: PR [W13] \+
C63: PR [W12] \+
D63: PR [W11] \+
E63: PR [W11] \+
P63: PR \-
Q63: PR \-
P64: PR 0
Q64: (F1) PR 2*@PI
A65: PR [W12] 'Adhesion Factors for Driven Piles (after Nordlund)
P65: PR 1
Q65: (F1) PR 7.6
A66: PR [W12] \-
B66: PR [W13] \-
C66: PR [W12] \-
D66: PR [W11] \-
E66: PR [W11] \-
F66: PR [W9] \-
G66: PR [W10] \-
P66: PR 2
Q66: (F1) PR 8.3
A67: PR [W12] "Geotechnical Model
D67: PR [W11] '|Model 1 :
E67: PR [W11] 'Sand/gravel over Stiff Clay
P67: PR 3
Q67: (F1) PR 8.8
A68: PR [W12] @IF(A36=4,"Not Applicable in this analysis",+A36)
D68: PR [W11] '|Model 2 :
E68: PR [W11] 'Soft Clay over Stiff Clay
P68: PR 4
Q68: (F1) PR 9
A69: PR [W12] @IF(A68=1,"Sand/Gravel thickness",@IF(A68=2,"Soft Clay thickness ","Upper
layer thickness = 0"))
D69: PR [W11] '|Model 3 :
E69: PR [W11] 'Only Stiff Clay
A70: PR [W12] +A38
B70: PR [W13] "metres
D70: PR [W11] '|Model 4 :
E70: PR [W11] 'Not applicable here
A71: PR [W12] \-
B71: PR [W13] \-
C71: PR [W12] \-
D71: PR [W11] \-
E71: PR [W11] \-
F71: PR [W9] \-
G71: PR [W10] \-
B72: PR [W13] "Undrained Shear
D72: PR [W11] 'Adhesion factor at depth embedded into stiff clay
B73: PR [W13] "Strength
D73: PR [W11] \-
E73: PR [W11] "-------------------------
G73: PR [W10] "------------------
B74: PR [W13] "Cu (kPa)

```
D74: PR [W11] " L <= 10 B
E74: PR [W11] "15 B
F74: PR [W9] "20 B
G74: PR [W10] "30 B
H74: PR [W9] "L > 40 B
B75: PR [W13] \-
C75: PR [W12] \-
D75: PR [W11] \-
E75: PR [W11] \-
F75: PR [W9] \-
G75: PR [W10] \-
H75: PR [W9] \-
B77: PR [W13] 50
D77: PR [W11] @IF($A$68=1,1,@IF($A$68=2,0.4,0.9))
E77: PR [W11] @IF($A$68=1,1,@IF($A$68=2,0.55,0.92))
F77: PR [W9] @IF($A$68=1,1,@IF($A$68=2,0.75,0.95))
G77: PR [W10] @IF($A$68=1,0.95,@IF($A$68=2,0.75,1))
H77: PR [W9] @IF($A$68=1,0.9,@IF($A$68=2,0.75,1))
B78: PR [W13] +C50
D78: PR [W11] @IF($A$68=1,1,@IF($A$68=2,0.3,0.65))
E78: PR [W11] @IF($A$68=1,0.95,@IF($A$68=2,0.5,0.67))
F78: PR [W9] @IF($A$68=1,0.9,@IF($A$68=2,0.69,0.7))
G78: PR [W10] @IF($A$68=1,0.75,@IF($A$68=2,0.69,0.75))
H78: PR [W9] @IF($A$68=1,0.7,@IF($A$68=2,0.69,0.85))
B79: PR [W13] 150
D79: PR [W11] @IF($A$68=1,1,@IF($A$68=2,0.25,0.3))
E79: PR [W11] @IF($A$68=1,0.9,@IF($A$68=2,0.45,0.32))
F79: PR [W9] @IF($A$68=1,0.75,@IF($A$68=2,0.63,0.35))
G79: PR [W10] @IF($A$68=1,0.55,@IF($A$68=2,0.63,0.4))
H79: PR [W9] @IF($A$68=1,0.35,@IF($A$68=2,0.63,0.45))
B80: PR [W13] 200
D80: PR [W11] @IF($A$68=1,1,@IF($A$68=2,0.2,0.25))
E80: PR [W11] @IF($A$68=1,0.9,@IF($A$68=2,0.4,0.26))
F80: PR [W9] @IF($A$68=1,0.75,@IF($A$68=2,0.55,0.27))
G80: PR [W10] @IF($A$68=1,0.55,@IF($A$68=2,0.55,0.28))
H80: PR [W9] @IF($A$68=1,0.35,@IF($A$68=2,0.55,0.3))
A82: PR [W12] \-
B82: PR [W13] \-
C82: PR [W12] \-
D82: PR [W11] \-
E82: PR [W11] \-
F82: PR [W9] \-
G82: PR [W10] \-
H82: PR [W9] \-
A83: PR [W12] 'Table is for a Geotechnical Model of
D83: PR [W11] @IF(A68=1,E67,@IF(A68=2,E68,E69))
A84: PR [W12] \-
B84: PR [W13] \-
C84: PR [W12] \-
D84: PR [W11] \-
E84: PR [W11] \-
F84: PR [W9] \-
G84: PR [W10] \-
H84: PR [W9] \-
I84: PR \-
J84: PR [W11] \-
B85: PR [W13] ^Depth
```

C85: PR [W12] "Undrained Shear
E85: PR [W11] 'Adhesion factor at depth embedded into stiff clay
A86: PR [W12] 'Depth
B86: PR [W13] ^(metres)
C86: PR [W12] "Strength
E86: PR [W11] "--------------------
G86: PR [W10] "--------------------
I86: PR \-
J86: PR [W11] \-
A87: PR [W12] 'Increment
C87: PR [W12] "Cu (kPa)
D87: (F2) PR [W11] 10*$F$18/1000
E87: (F2) PR [W11] 15*$F$18/1000
F87: (F2) PR [W9] 20*$F$18/1000
G87: (F2) PR [W10] 30*$F$18/1000
H87: (F2) PR [W9] 40*$F$18/1000
I87: (F1) PR 'Design Value
A88: PR [W12] \-
B88: PR [W13] \-
C88: PR [W12] \-
D88: PR [W11] \-
E88: PR [W11] \-
F88: PR [W9] \-
G88: PR [W10] \-
H88: PR [W9] \-
I88: PR \-
J88: PR [W11] \-
A89: PR [W12] 0
B89: PR [W13] +A48
C89: PR [W12] +C48
D89: PR [W11] @IF(B89<=$A$70,0,@VLOOKUP(C89,$B$77..$H$80,2))
E89: PR [W11] @IF(B89<=$A$70,0,@VLOOKUP(C89,$B$77..$H$80,3))
F89: PR [W9] @IF(B89<=$A$70,0,@VLOOKUP(C89,$B$77..$H$80,4))
G89: PR [W10] @IF(B89<=$A$70,0,@VLOOKUP(C89,$B$77..$H$80,5))
H89: PR [W9] @IF(B89<=$A$70,0,@VLOOKUP(C89,$B$77..$H$80,6))
I89: PR @IF((B89-$A$70)>=$D$87,@HLOOKUP((B89-$A$70),D$87..$H$99,(+A89+1)),+D89)
A90: PR [W12] +A89+1
B90: PR [W13] +A49
C90: PR [W12] +C49
D90: PR [W11] @IF(B90<=$A$70,0,@VLOOKUP(C90,$B$77..$H$80,2))
E90: PR [W11] @IF(B90<=$A$70,0,@VLOOKUP(C90,$B$77..$H$80,3))
F90: PR [W9] @IF(B90<=$A$70,0,@VLOOKUP(C90,$B$77..$H$80,4))
G90: PR [W10] @IF(B90<=$A$70,0,@VLOOKUP(C90,$B$77..$H$80,5))
H90: PR [W9] @IF(B90<=$A$70,0,@VLOOKUP(C90,$B$77..$H$80,6))
I90: PR @IF((B90-$A$70)>=$D$87,@HLOOKUP((B90-$A$70),D$87..$H$99,(+A90+1)),+D90)
A91: PR [W12] +A90+1
B91: PR [W13] +A50
C91: PR [W12] +C50
D91: PR [W11] @IF(B91<=$A$70,0,@VLOOKUP(C91,$B$77..$H$80,2))
E91: PR [W11] @IF(B91<=$A$70,0,@VLOOKUP(C91,$B$77..$H$80,3))
F91: PR [W9] @IF(B91<=$A$70,0,@VLOOKUP(C91,$B$77..$H$80,4))
G91: PR [W10] @IF(B91<=$A$70,0,@VLOOKUP(C91,$B$77..$H$80,5))
H91: PR [W9] @IF(B91<=$A$70,0,@VLOOKUP(C91,$B$77..$H$80,6))
I91: PR @IF((B91-$A$70)>=$D$87,@HLOOKUP((B91-$A$70),D$87..$H$99,(+A91+1)),+D91)
A92: PR [W12] +A91+1
B92: PR [W13] +A51
C92: PR [W12] +C51

D92: PR [W11] @IF(B92<=$A$70,0,@VLOOKUP(C92,$B$77..$H$80,2))
E92: PR [W11] @IF(B92<=$A$70,0,@VLOOKUP(C92,$B$77..$H$80,3))
F92: PR [W9] @IF(B92<=$A$70,0,@VLOOKUP(C92,$B$77..$H$80,4))
G92: PR [W10] @IF(B92<=$A$70,0,@VLOOKUP(C92,$B$77..$H$80,5))
H92: PR [W9] @IF(B92<=$A$70,0,@VLOOKUP(C92,$B$77..$H$80,6))
I92: PR @IF((B92-$A$70)>=$D$87,@HLOOKUP((B92-$A$70),D$87..$H$99,(+A92+1)),+D92)
A93: PR [W12] +A92+1
B93: PR [W13] +A52
C93: PR [W12] +C52
D93: PR [W11] @IF(B93<=$A$70,0,@VLOOKUP(C93,$B$77..$H$80,2))
E93: PR [W11] @IF(B93<=$A$70,0,@VLOOKUP(C93,$B$77..$H$80,3))
F93: PR [W9] @IF(B93<=$A$70,0,@VLOOKUP(C93,$B$77..$H$80,4))
G93: PR [W10] @IF(B93<=$A$70,0,@VLOOKUP(C93,$B$77..$H$80,5))
H93: PR [W9] @IF(B93<=$A$70,0,@VLOOKUP(C93,$B$77..$H$80,6))
I93: PR @IF((B93-$A$70)>=$D$87,@HLOOKUP((B93-$A$70),D$87..$H$99,(+A93+1)),+D93)
A94: PR [W12] +A93+1
B94: PR [W13] +A53
C94: PR [W12] +C53
D94: PR [W11] @IF(B94<=$A$70,0,@VLOOKUP(C94,$B$77..$H$80,2))
E94: PR [W11] @IF(B94<=$A$70,0,@VLOOKUP(C94,$B$77..$H$80,3))
F94: PR [W9] @IF(B94<=$A$70,0,@VLOOKUP(C94,$B$77..$H$80,4))
G94: PR [W10] @IF(B94<=$A$70,0,@VLOOKUP(C94,$B$77..$H$80,5))
H94: PR [W9] @IF(B94<=$A$70,0,@VLOOKUP(C94,$B$77..$H$80,6))
I94: PR @IF((B94-$A$70)>=$D$87,@HLOOKUP((B94-$A$70),D$87..$H$99,(+A94+1)),+D94)
A95: PR [W12] +A94+1
B95: PR [W13] +A54
C95: PR [W12] +C54
D95: PR [W11] @IF(B95<=$A$70,0,@VLOOKUP(C95,$B$77..$H$80,2))
E95: PR [W11] @IF(B95<=$A$70,0,@VLOOKUP(C95,$B$77..$H$80,3))
F95: PR [W9] @IF(B95<=$A$70,0,@VLOOKUP(C95,$B$77..$H$80,4))
G95: PR [W10] @IF(B95<=$A$70,0,@VLOOKUP(C95,$B$77..$H$80,5))
H95: PR [W9] @IF(B95<=$A$70,0,@VLOOKUP(C95,$B$77..$H$80,6))
I95: PR @IF((B95-$A$70)>=$D$87,@HLOOKUP((B95-$A$70),D$87..$H$99,(+A95+1)),+D95)
A96: PR [W12] +A95+1
B96: PR [W13] +A55
C96: PR [W12] +C55
D96: PR [W11] @IF(B96<=$A$70,0,@VLOOKUP(C96,$B$77..$H$80,2))
E96: PR [W11] @IF(B96<=$A$70,0,@VLOOKUP(C96,$B$77..$H$80,3))
F96: PR [W9] @IF(B96<=$A$70,0,@VLOOKUP(C96,$B$77..$H$80,4))
G96: PR [W10] @IF(B96<=$A$70,0,@VLOOKUP(C96,$B$77..$H$80,5))
H96: PR [W9] @IF(B96<=$A$70,0,@VLOOKUP(C96,$B$77..$H$80,6))
I96: PR @IF((B96-$A$70)>=$D$87,@HLOOKUP((B96-$A$70),D$87..$H$99,(+A96+1)),+D96)
A97: PR [W12] +A96+1
B97: PR [W13] +A56
C97: PR [W12] +C56
D97: PR [W11] @IF(B97<=$A$70,0,@VLOOKUP(C97,$B$77..$H$80,2))
E97: PR [W11] @IF(B97<=$A$70,0,@VLOOKUP(C97,$B$77..$H$80,3))
F97: PR [W9] @IF(B97<=$A$70,0,@VLOOKUP(C97,$B$77..$H$80,4))
G97: PR [W10] @IF(B97<=$A$70,0,@VLOOKUP(C97,$B$77..$H$80,5))
H97: PR [W9] @IF(B97<=$A$70,0,@VLOOKUP(C97,$B$77..$H$80,6))
I97: PR @IF((B97-$A$70)>=$D$87,@HLOOKUP((B97-$A$70),D$87..$H$99,(+A97+1)),+D97)
A98: PR [W12] +A97+1
B98: PR [W13] +A57
C98: PR [W12] +C57
D98: PR [W11] @IF(B98<=$A$70,0,@VLOOKUP(C98,$B$77..$H$80,2))
E98: PR [W11] @IF(B98<=$A$70,0,@VLOOKUP(C98,$B$77..$H$80,3))
F98: PR [W9] @IF(B98<=$A$70,0,@VLOOKUP(C98,$B$77..$H$80,4))

```
G98: PR [W10] @IF(B98<=$A$70,0,@VLOOKUP(C98,$B$77..$H$80,5))
H98: PR [W9] @IF(B98<=$A$70,0,@VLOOKUP(C98,$B$77..$H$80,6))
I98: PR @IF((B98-$A$70)>=$D$87,@HLOOKUP((B98-$A$70),D$87..$H$99,(+A98+1)),+D98)
A99: PR [W12] +A98+1
B99: PR [W13] +A58
C99: PR [W12] +C58
D99: PR [W11] @IF(B99<=$A$70,0,@VLOOKUP(C99,$B$77..$H$80,2))
E99: PR [W11] @IF(B99<=$A$70,0,@VLOOKUP(C99,$B$77..$H$80,3))
F99: PR [W9] @IF(B99<=$A$70,0,@VLOOKUP(C99,$B$77..$H$80,4))
G99: PR [W10] @IF(B99<=$A$70,0,@VLOOKUP(C99,$B$77..$H$80,5))
H99: PR [W9] @IF(B99<=$A$70,0,@VLOOKUP(C99,$B$77..$H$80,6))
I99: PR @IF((B99-$A$70)>=$D$87,@HLOOKUP((B99-$A$70),D$87..$H$99,(+A99+1)),+D99)
A100: PR [W12] \-
B100: PR [W13] \-
C100: PR [W12] \-
D100: PR [W11] \-
E100: PR [W11] \-
F100: PR [W9] \-
G100: PR [W10] \-
H100: PR [W9] \-
I100: PR \-


10. FILE: FOUNDBRG

B1: PR \:
C1: PR '   Spread Footing Design
F1: PR \:
B2: PR \:
C2: PR \:
D2: PR \:
E2: PR \:
F2: PR \:
B4: PR ' Soil Properties
E4: PR "Value
F4: PR 'Units
H4: PR [W10] ' Description
A5: PR [W2] "*
B5: PR \*
C5: PR \*
D5: PR \*
E5: PR \*
F5: PR \*
G5: PR [W2] '*
H5: PR [W10] \-
I5: PR \-
A6: PR [W2] "*
C6: PR '      Cohesion ?
E6: U 0
F6: PR 'kPa
G6: PR [W2] '*
J6: PR [W2] "|
A7: PR [W2] "*
B7: PR '       Angle of Friction ?
E7: U 30
F7: PR @CHAR(176)
```

G7: PR [W2] '*

H7: (F4) PR [W10] +E7/180*@PI

I7: PR 'radians

J7: PR [W2] "|

A8: PR [W2] "*

C8: PR 'Bulk Unit Weight ?

E8: U 18

F8: PR 'kN/ cu m

G8: PR [W2] '*

J8: PR [W2] "|

A9: PR [W2] "*

B9: PR '   Saturated Unit Weight ?

E9: U 18

F9: PR 'kN/ cu m

G9: PR [W2] '*

J9: PR [W2] "|

A10: PR [W2] "*

B10: PR *

C10: PR *

D10: PR *

E10: PR *

F10: PR *

G10: PR [W2] '*

J10: PR [W2] "|

J11: PR [W2] "|

B12: PR 'Proposed Footing Geometry

J12: PR [W2] "|

A13: PR [W2] "*

B13: PR *

C13: PR *

D13: PR *

E13: PR *

F13: PR *

G13: PR [W2] '*

J13: PR [W2] "|

A14: PR [W2] "*

C14: PR 'Width of Footing ?

E14: (F3) PR +H53

F14: PR 'metres

G14: PR [W2] '*

J14: PR [W2] "|

A15: PR [W2] "*

B15: PR 'Footing length, (D) or (S)?

E15: U 's

F15: PR 'metres

G15: PR [W2] '*

H15: PR [W10] '(D)iamater:(S)quare

J15: PR [W2] "|

A16: PR [W2] "*

B16: PR '   Depth of Footing (Df) ?

E16: U 0.7

F16: PR 'metres

G16: PR [W2] '*

H16: (F2) PR [W10] @IF(@ROUND(E34,1)=@ROUND(H51,1),"Design size",@IF(@ROUND
(E34,1)>@ROUND(H51,1),"Footing size unsuitable","Footing size suitable, but"))

J16: PR [W2] "|

A17: PR [W2] "*

B17: PR '   Thickness of Footing ?
E17: U 0.3
F17: PR 'metres
G17: PR [W2] '*
H17: (F2) PR [W10] @IF(@ROUND(E34,1)=@ROUND(H51,1),"Required",@IF(@ROUND
(E34,1)>@ROUND(H51,1),"Adjust size","Check for economical size"))
J17: PR [W2] "|
A18: PR [W2] "*
B18: PR *
C18: PR *
D18: PR *
E18: PR *
F18: PR *
G18: PR [W2] '*
J18: PR [W2] "|
J19: PR [W2] "|
B20: PR 'Loading Conditions
J20: PR [W2] "|
A21: PR [W2] "*
B21: PR *
C21: PR *
D21: PR *
E21: PR *
F21: PR *
G21: PR [W2] '*
J21: PR [W2] "|
A22: PR [W2] "*
B22: PR '   Depth of Groundwater ?
E22: U 3
F22: PR 'metres
G22: PR [W2] '*
H22: PR [W10] 'below surface level
J22: PR [W2] "|
A23: PR [W2] "*
B23: PR '   Applied column load ?
E23: U 150
F23: PR 'kN
G23: PR [W2] '*
H23: PR [W10] 'at top of footing
J23: PR [W2] "|
A24: PR [W2] "*
B24: PR ' Load Inclination {alpha} ?
E24: U 20
F24: PR @CHAR(176)
G24: PR [W2] '*
H24: (F4) PR [W10] +E24/180*@PI
I24: PR 'radians
J24: PR [W2] "|
A25: PR [W2] "*
B25: PR ' Unit Weight of concrete ?
E25: U 24
F25: PR 'kN /cu m
G25: PR [W2] '*
J25: PR [W2] "|
A26: PR [W2] "*
C26: PR 'Factor of safety ?
E26: U 3

G26: PR [W2] '*
J26: PR [W2] "|
A27: PR [W2] "*
B27: PR *
C27: PR *
D27: PR *
E27: PR *
F27: PR *
J27: PR [W2] "|
J28: PR [W2] "|
B29: PR 'Calculated Quantities
J29: PR [W2] "|
A30: PR [W2] "*
B30: PR *
C30: PR *
D30: PR *
E30: PR *
F30: PR *
G30: PR [W2] '*
J30: PR [W2] "|
A31: PR [W2] "*
B31: PR 'Vol of concrete in footing
E31: (F3) PR @IF(E15="d",@PI*E14^2*E17/4,@IF(E15="s",E14^2*E17,E14*E15*E17))
F31: PR 'cu metre
G31: PR [W2] '*
J31: PR [W2] "|
A32: PR [W2] "*
B32: PR '==> Weight of Footing =
E32: (F1) PR +E31*E25
F32: PR 'kN
G32: PR [W2] '*
J32: PR [W2] "|
A33: PR [W2] "*
B33: PR 'Weight of equivalent soil
E33: (F1) PR @IF(E22>E16,E31*E8,E31*(E9-9.81))
F33: PR 'kN
G33: PR [W2] '*
J33: PR [W2] "|
A34: PR [W2] "*
B34: PR '== > Net applied load =
E34: (F1) PR +E23+E32-E33
F34: PR 'kN
G34: PR [W2] '*
J34: PR [W2] "|
A35: PR [W2] "*
B35: PR 'Initial overburden pressure
E35: (F2) PR @IF(E22>E16,E8*E16,(E8*E22)+(E16-E22)*(E9-9.81))
F35: PR 'kPa
G35: PR [W2] '*
H35: PR [W10] 'at base of footing
J35: PR [W2] "|
A36: PR [W2] "*
B36: PR *
C36: PR *
D36: PR *
E36: PR *
F36: PR *

G36: PR [W2] '*
J36: PR [W2] "|
J37: PR [W2] "|
B38: PR 'Calculation Table of Bearing Capacity factors
J38: PR [W2] "|
A39: PR [W2] "+
B39: PR \+
C39: PR \+
D39: PR \+
E39: PR \+
F39: PR \+
G39: PR [W2] \+
H39: PR [W10] \+
I39: PR \+
J39: PR [W2] '+
A40: PR [W2] "+
C40: PR 'Brg Capac
D40: PR '   Geometry & Load factors
H40: PR [W10] ' Ultimate Bearing
J40: PR [W2] '+
A41: PR [W2] "+
C41: PR 'Factors
D41: PR "Shape
E41: PR ^Depth
F41: PR 'Inclination
H41: PR [W10] '   Capacity (kPa)
J41: PR [W2] '+
A42: PR [W2] "+
B42: PR \+
C42: PR \+
D42: PR \+
E42: PR \+
F42: PR \+
G42: PR [W2] \+
H42: PR [W10] \+
I42: PR \+
J42: PR [W2] '+
A43: PR [W2] "+
B43: PR 'q
C43: (F2) PR @EXP(@PI*@TAN(H7))*(@TAN(45/180*@PI+H7/2))^2
D43: (F2) PR @IF(E11="d"#OR#E11="s",1+@TAN(H7),1+(E14/E15)*@TAN(H7))
E43: (F2) PR @IF(E16/E14<=1,1+2*@TAN(H7)*(1-@SIN(H7))^2*E16/E14,1+2*@TAN(H7)*(1-@
SIN(H7))^2*@ATAN(E16/E14))
F43: (F2) PR (1-E24/90)^2
H43: (F2) PR [W10] +C43*D43*E43*F43*E35
J43: PR [W2] '+
A44: PR [W2] "+
B44: PR 'c
C44: (F2) PR @IF(H7=0,5.14,(C43-1)/@TAN(H7))
D44: (F2) PR @IF(E15="s"#OR#E15="d",1+C43/C44,1+(E14/E15)*(C43/C44))
E44: (F2) PR @IF(E16/E14<=1#AND#E7=0,1+0.4*E16/E14,@IF(E16/E14<=1#AND#E7>0,+E43-
(1-E43)/C43/@TAN(H7),@IF(E7=0#AND#E16/E14>1,1+0.4*@ATAN(E16/E14),+E43-(1-E43)/C43/@
TAN(H7))))
F44: (F2) PR (1-E24/90)^2
H44: (F2) PR [W10] +C44*D44*E44*F44*E6
J44: PR [W2] '+
A45: PR [W2] "+

B45: PR '{gamma}
C45: (F2) PR 2*(C43+1)*@TAN(H7)
D45: (F2) PR @IF(E15="d"#OR#E15="s",0.6,1-0.4*E14/E15)
E45: (F2) PR 1
F45: (F2) PR @IF(E7=0,1,(1-E24/E7)^2)
H45: (F2) PR [W10] @IF(E22>E16,+C45*D45*E45*F45*E8*E14/2,+C45*D45*E45*F45*(E9-9.81)
*E14/2)
J45: PR [W2] '+
A46: PR [W2] "+
J46: PR [W2] '+
A47: PR [W2] "+
B47: PR \+
C47: PR \+
D47: PR \+
E47: PR \+
F47: PR \+
G47: PR [W2] \+
H47: PR [W10] \+
I47: PR \+
J47: PR [W2] '+
B48: PR "+
C48: PR '   Total Ultimate Bearing capacity
H48: (F2) PR [W10] @SUM(H43..H45)
I48: PR 'kPa
J48: PR [W2] '+
B49: PR "+
D49: PR 'Allowable Bearing capacity
H49: (F2) PR [W10] +H48/E26
I49: PR 'kPa
J49: PR [W2] '+
B50: PR "+
D50: PR 'Allowable Total gross load
H50: (F1) PR [W10] @IF(E15="d",@PI/4*E14^2*H49,@IF(E15="s",E14^2*H49,E14*E15*H49))
I50: PR 'kN
J50: PR [W2] '+
B51: PR "+
D51: PR 'Allowable Total net load =
H51: (F1) PR [W10] @IF(E15="d",(H48-E35)/E26*(@PI/4*E14^2),@IF(E15="s",(H48-E35)/E26*
(E14^2),(H48-E35)/E26*(E14*E15)))
I51: PR 'kN
J51: PR [W2] '+
B52: PR "+
C52: PR \+
D52: PR \+
E52: PR \+
F52: PR \+
G52: PR [W2] \+
H52: PR [W10] \+
I52: PR \+
J52: PR [W2] '+
B53: PR "=
C53: PR @IF($E$15="d","Diameter of Footing required","Width of Footing (B) required")
H53: (F3) PR [W10] @IF(@ROUND(E34,1)=@ROUND(H51,1),E14,@IF(@ROUND(E34,1)>@
ROUND(H51,1),E14+0.001,E14-0.001))
I53: PR 'metres
J53: PR [W2] '=
B54: PR "=

C54: PR \=
D54: PR \=
E54: PR \=
F54: PR \=
G54: PR [W2] \=
H54: PR [W10] \=
I54: PR \=
J54: PR [W2] '=

11. FILE: FOUNDSET

A1: PR [W7] 'Settlements in granular Soil
A2: PR [W7] "----------------------------
A5: PR [W7] 'Proposed Foundation type (R)aft or (F)ooting) ?
F5: U [W8] ^f
G5: PR [W10] @IF(F5="R","Raft","Footing")
C6: PR [W11] @IF($F$5="R","Proposed Raft size B","Proposed Footing size B")
F6: U [W8] 1.5
G6: PR [W10] ^metres
C7: PR [W11] ^Applied Foundation Pressure
F7: U [W8] 100
G7: PR [W10] ^kPa
C8: PR [W11] ^Unit weight of soil
F8: U [W8] 20
G8: PR [W10] ^kN / cu m
C9: PR [W11] ^Depth to water table
F9: U [W8] 10
G9: PR [W10] ^metres
A10: PR [W7] 'Is the soil below the water Fine or Silty sand ?
F10: U [W8] ^Y
G10: PR [W10] '(Y)es or (N)o
A11: PR [W7] "+++++++++++++
B11: PR [W12] "++ı ı ı +++++++++++
C11: PR [W11] "++++++++++++++++
D11: PR [W11] "++++++++++++++
E11: PR [W10] "+++++++++++++
F11: PR [W8] "+++++++++++++
G11: PR [W10] "+++++++++++++
H11: PR "+++++++++++++
B12: PR [W12] ^S.P.T.
C12: PR [W11] ^Effective
D12: PR [W11] ^Corrected N Value
F12: PR [W8] ^Applied
G12: PR [W10] 'SETTLEMENT mm
I12: (H) PR ^PIE CHART
A13: PR [W7] 'Depth
B13: PR [W12] ^N Value
C13: PR [W11] ^Overburden
D13: PR [W11] ^For
E13: PR [W10] ^For
F13: PR [W8] ^Stress
G13: PR [W10] ^at depth
H13: PR ^Total
I13: (H) PR ^ENHANCEMENTS
A14: PR [W7] ^metres

B14: PR [W12] ^Uncorrected
C14: PR [W11] ^Press. kPa
D14: PR [W11] ^Overburden
E14: PR [W10] ^Saturation
F14: PR [W8] ^kPa
G14: PR [W10] ^increment
H14: PR ^at depth
A15: PR [W7] "+++++++++++++
B15: PR [W12] "+++++++++++++++
C15: PR [W11] "+++++++++++++++++
D15: PR [W11] "+++++++++++++++
E15: PR [W10] "+++++++++++++
F15: PR [W8] "+++++++++++++
G15: PR [W10] "+++++++++++++
H15: PR "+++++++++++++
A16: U [W7] 1.5
B16: U [W12] 8
C16: (F2) PR [W11] @IF(A16<=$F$9,+$F$8*A16,($F$8-9.8)*(A16-$F$9)+$F$8*$F$9)
D16: (F0) PR [W11] @IF(C16>23.9,0.77*B16*@LOG(20/0.0105/C16),@IF(C16>=15,1.6,@IF
(C16>=6,1.8,0)))
E16: (F0) PR [W10] @IF(A16>$F$9,15+0.5*(D16-15),D16)
F16: (F2) PR [W8] +$F$7/@PI*(@ATAN($F$6/A16)+@SIN(@ATAN($F$6/A16)))
G16: (F1) PR [W10] @IF($F$5="r",2.84*F16/E16,@IF($F$6>1.25,2.84*F16/E16*($F$6/($F$6+
0.33))^2,1.9*F16/E16))
H16: (F1) PR @SUM($G$16..G16)
I16: (H) PR @IF(G16>0.1*@SUM($G$16..$G$25),1,@IF(G16<0.05*@SUM($G$16..$G$25),
100,0))
A17: U [W7] 3
B17: U [W12] 12
C17: (F2) PR [W11] @IF(A17<=$F$9,+$F$8*A17,($F$8-9.8)*(A17-$F$9)+$F$8*$F$9)
D17: (F0) PR [W11] @IF(C17>23.9,0.77*B17*@LOG(20/0.0105/C17),@IF(C17>=15,1.6,@IF
(C17>=6,1.8,0)))
E17: (F0) PR [W10] @IF(A17>$F$9,15+0.5*(D17-15),D17)
F17: (F2) PR [W8] +$F$7/@PI*(@ATAN($F$6/A17)+@SIN(@ATAN($F$6/A17)))
G17: (F1) PR [W10] @IF($F$5="r",2.84*F17/E17,@IF($F$6>1.25,2.84*F17/E17*($F$6/($F$6+
0.33))^2,1.9*F17/E17))
H17: (F1) PR @SUM($G$16..G17)
I17: (H) PR @IF(G17>0.1*@SUM($G$16..$G$25),1,@IF(G17<0.05*@SUM($G$16..$G$25),
100,0))
A18: U [W7] 4.5
B18: U [W12] 15
C18: (F2) PR [W11] @IF(A18<=$F$9,+$F$8*A18,($F$8-9.8)*(A18-$F$9)+$F$8*$F$9)
D18: (F0) PR [W11] @IF(C18>23.9,0.77*B18*@LOG(20/0.0105/C18),@IF(C18>=15,1.6,@IF
(C18>=6,1.8,0)))
E18: (F0) PR [W10] @IF(A18>$F$9,15+0.5*(D18-15),D18)
F18: (F2) PR [W8] +$F$7/@PI*(@ATAN($F$6/A18)+@SIN(@ATAN($F$6/A18)))
G18: (F1) PR [W10] @IF($F$5="r",2.84*F18/E18,@IF($F$6>1.25,2.84*F18/E18*($F$6/($F$6+
0.33))^2,1.9*F18/E18))
H18: (F1) PR @SUM($G$16..G18)
I18: (H) PR @IF(G18>0.1*@SUM($G$16..$G$25),1,@IF(G18<0.05*@SUM($G$16..$G$25),
100,0))
A19: U [W7] 6
B19: U [W12] 9
C19: (F2) PR [W11] @IF(A19<=$F$9,+$F$8*A19,($F$8-9.8)*(A19-$F$9)+$F$8*$F$9)
D19: (F0) PR [W11] @IF(C19>23.9,0.77*B19*@LOG(20/0.0105/C19),@IF(C19>=15,1.6,@IF
(C19>=6,1.8,0)))
E19: (F0) PR [W10] @IF(A19>$F$9,15+0.5*(D19-15),D19)

F19: (F2) PR [W8] +$F$7/@PI*(@ATAN($F$6/A19)+@SIN(@ATAN($F$6/A19)))

G19: (F1) PR [W10] @IF($F$5="r",2.84*F19/E19,@IF($F$6>1.25,2.84*F19/E19*($F$6/($F$6+
0.33))^2,1.9*F19/E19))

H19: (F1) PR @SUM($G$16..G19)

I19: (H) PR @IF(G19>0.1*@SUM($G$16..$G$25),1,@IF(G19<0.05*@SUM($G$16..$G$25),
100,0))

A20: U [W7] 7.5

B20: U [W12] 7

C20: (F2) PR [W11] @IF(A20<=$F$9,+$F$8*A20,($F$8-9.8)*(A20-$F$9)+$F$8*$F$9)

D20: (F0) PR [W11] @IF(C20>23.9,0.77*B20*@LOG(20/0.0105/C20),@IF(C20>=15,1.6,@IF
(C20>=6,1.8,0)))

E20: (F0) PR [W10] @IF(A20>$F$9,15+0.5*(D20-15),D20)

F20: (F2) PR [W8] +$F$7/@PI*(@ATAN($F$6/A20)+@SIN(@ATAN($F$6/A20)))

G20: (F1) PR [W10] @IF($F$5="r",2.84*F20/E20,@IF($F$6>1.25,2.84*F20/E20*($F$6/($F$6+
0.33))^2,1.9*F20/E20))

H20: (F1) PR @SUM($G$16..G20)

I20: (H) PR @IF(G20>0.1*@SUM($G$16..$G$25),1,@IF(G20<0.05*@SUM($G$16..$G$25),
100,0))

A21: U [W7] 9

B21: U [W12] 12

C21: (F2) PR [W11] @IF(A21<=$F$9,+$F$8*A21,($F$8-9.8)*(A21-$F$9)+$F$8*$F$9)

D21: (F0) PR [W11] @IF(C21>23.9,0.77*B21*@LOG(20/0.0105/C21),@IF(C21>=15,1.6,@IF
(C21>=6,1.8,0)))

E21: (F0) PR [W10] @IF(A21>$F$9,15+0.5*(D21-15),D21)

F21: (F2) PR [W8] +$F$7/@PI*(@ATAN($F$6/A21)+@SIN(@ATAN($F$6/A21)))

G21: (F1) PR [W10] @IF($F$5="r",2.84*F21/E21,@IF($F$6>1.25,2.84*F21/E21*($F$6/($F$6+
0.33))^2,1.9*F21/E21))

H21: (F1) PR @SUM($G$16..G21)

I21: (H) PR @IF(G21>0.1*@SUM($G$16..$G$25),1,@IF(G21<0.05*@SUM($G$16..$G$25),
100,0))

A22: U [W7] 10.5

B22: U [W12] 15

C22: (F2) PR [W11] @IF(A22<=$F$9,+$F$8*A22,($F$8-9.8)*(A22-$F$9)+$F$8*$F$9)

D22: (F0) PR [W11] @IF(C22>23.9,0.77*B22*@LOG(20/0.0105/C22),@IF(C22>=15,1.6,@IF
(C22>=6,1.8,0)))

E22: (F0) PR [W10] @IF(A22>$F$9,15+0.5*(D22-15),D22)

F22: (F2) PR [W8] +$F$7/@PI*(@ATAN($F$6/A22)+@SIN(@ATAN($F$6/A22)))

G22: (F1) PR [W10] @IF($F$5="r",2.84*F22/E22,@IF($F$6>1.25,2.84*F22/E22*($F$6/($F$6+
0.33))^2,1.9*F22/E22))

H22: (F1) PR @SUM($G$16..G22)

I22: (H) PR @IF(G22>0.1*@SUM($G$16..$G$25),1,@IF(G22<0.05*@SUM($G$16..$G$25),
100,0))

A23: U [W7] 12

B23: U [W12] 13

C23: (F2) PR [W11] @IF(A23<=$F$9,+$F$8*A23,($F$8-9.8)*(A23-$F$9)+$F$8*$F$9)

D23: (F0) PR [W11] @IF(C23>23.9,0.77*B23*@LOG(20/0.0105/C23),@IF(C23>=15,1.6,@IF
(C23>=6,1.8,0)))

E23: (F0) PR [W10] @IF(A23>$F$9,15+0.5*(D23-15),D23)

F23: (F2) PR [W8] +$F$7/@PI*(@ATAN($F$6/A23)+@SIN(@ATAN($F$6/A23)))

G23: (F1) PR [W10] @IF($F$5="r",2.84*F23/E23,@IF($F$6>1.25,2.84*F23/E23*($F$6/($F$6+
0.33))^2,1.9*F23/E23))

H23: (F1) PR @SUM($G$16..G23)

I23: (H) PR @IF(G23>0.1*@SUM($G$16..$G$25),1,@IF(G23<0.05*@SUM($G$16..$G$25),
100,0))

A24: U [W7] 13.5

B24: U [W12] 20

C24: (F2) PR [W11] @IF(A24<=$F$9,+$F$8*A24,($F$8-9.8)*(A24-$F$9)+$F$8*$F$9)

D24: (F0) PR [W11] @IF(C24>23.9,0.77*B24*@LOG(20/0.0105/C24),@IF(C24>=15,1.6,@IF
(C24>=6,1.8,0)))

E24: (F0) PR [W10] @IF(A24>$F$9,15+0.5*(D24-15),D24)

F24: (F2) PR [W8] +$F$7/@PI*(@ATAN($F$6/A24)+@SIN(@ATAN($F$6/A24)))

G24: (F1) PR [W10] @IF($F$5="r",2.84*F24/E24,@IF($F$6>1.25,2.84*F24/E24*($F$6/($F$6+
0.33))^2,1.9*F24/E24))

H24: (F1) PR @SUM($G$16..G24)

I24: (H) PR @IF(G24>0.1*@SUM($G$16..$G$25),1,@IF(G24<0.05*@SUM($G$16..$G$25),
100,0))

A25: U [W7] 15

B25: U [W12] 25

C25: (F2) PR [W11] @IF(A25<=$F$9,+$F$8*A25,($F$8-9.8)*(A25-$F$9)+$F$8*$F$9)

D25: (F0) PR [W11] @IF(C25>23.9,0.77*B25*@LOG(20/0.0105/C25),@IF(C25>=15,1.6,@IF
(C25>=6,1.8,0)))

E25: (F0) PR [W10] @IF(A25>$F$9,15+0.5*(D25-15),D25)

F25: (F2) PR [W8] +$F$7/@PI*(@ATAN($F$6/A25)+@SIN(@ATAN($F$6/A25)))

G25: (F1) PR [W10] @IF($F$5="r",2.84*F25/E25,@IF($F$6>1.25,2.84*F25/E25*($F$6/($F$6+
0.33))^2,1.9*F25/E25))

H25: (F1) PR @SUM($G$16..G25)

I25: (H) PR @IF(G25>0.1*@SUM($G$16..$G$25),1,@IF(G25<0.05*@SUM($G$16..$G$25),
100,0))

A26: PR [W7] "+++++++++++++

B26: PR [W12] "+++++++++++++++

C26: PR [W11] "+++++++++++++++

D26: PR [W11] "+++++++++++++

E26: PR [W10] "+++++++++++++

F26: PR [W8] "+++++++++++++

G26: PR [W10] "+++++++++++++

H26: PR "+++++++++++++

12. FILE: EMBANK1

A1: PR [W17] 'CONSOLIDATION WITH TIME, FILL HEIGHT & VERTICAL DRAINS

A2: PR [W17] \-

B2: PR [W12] \-

C2: PR \-

D2: PR \-

E2: PR \-

B3: PR [W12] 'JOB TITLE :

D3: U 'Coomera

B5: PR [W12] 'SOIL AND SURCHARGE PROPERTIES

A6: PR [W17] \+

B6: PR [W12] \+

C6: PR \+

D6: PR \+

E6: PR \+

F6: PR '+

A7: PR [W17] 'Coefficient of compressibility Mv

D7: U 0.001

E7: PR 'sq m /kN

F7: PR '+

A8: PR [W17] 'Coeff. of Consolidation (Vertical) Cv

D8: U 5

E8: PR 'sq m /yr

F8: PR '+

A9: PR [W17] 'Coeff. of Consolidation (Radial) Cvr
D9: U 10
E9: PR 'sq m /yr
F9: PR '+
A10: PR [W17] 'Thickness of compressible clay layer H
D10: U 5
E10: PR 'metres
F10: PR '+
A11: PR [W17] 'Unit weight of placed fill
D11: U 18
E11: PR 'kN/cu m
F11: PR '+
A12: PR [W17] 'Design Height of Fill
D12: U 2.5
E12: PR 'metres
F12: PR '+
A13: PR [W17] 'Increment Fill Height
D13: U 1
E13: PR 'metres
F13: PR '+
F14: PR '+
B15: PR [W12] 'PROPERTIES OF VERTICAL DRAINS
F15: PR '+
A16: PR [W17] \+
B16: PR [W12] \+
C16: PR \+
D16: PR \+
E16: PR \+
F16: PR '+
B17: PR [W12] 'Drain Radius rd
D17: U 0.15
E17: PR 'metres
F17: PR '+
B18: PR [W12] 'Spacing of Drains S
D18: U 3
E18: PR 'metres
F18: PR '+
A19: PR [W17] '(S)quare or (T)riangular pattern
D19: U 's
F19: PR '+
B20: PR [W12] '==> Equivalent radius (R) =
E20: PR @IF(D19="s",0.564*D18,0.525*D18)
F20: PR '+
A21: PR [W17] 'Factor to account for smear :
C21: U 1.5
D21: PR '* R
E21: PR +E20*C21
F21: PR '+
C22: PR "======>
D22: PR "n =
E22: PR +E21/2/D17
F22: PR '+
C23: PR "======>
D23: PR "m =
E23: PR +E22^2/(E22^2-1)*@LN(E22)-(3*E22^2-1)/4/E22^2
F23: PR '+
F24: PR '+

C25: PR 'TIME FRAMES
F25: PR '+
A26: PR [W17] \+
B26: PR [W12] \+
C26: PR \+
D26: PR \+
E26: PR \+
F26: PR '+
B27: PR [W12] 'Start of time t0
D27: U 0
E27: PR 'weeks
F27: PR '+
B28: PR [W12] 'End of time t1
D28: U 40
E28: PR 'weeks
F28: PR '+
A29: PR [W17] \+
B29: PR [W12] \+
C29: PR \+
D29: PR \+
E29: PR \+
F29: PR \+
G29: PR \+
H29: PR [W2] '+
B30: PR [W12] 'Time (weeks)
C30: PR +D27
D30: PR +$C$30+($D$28-$D$27)/4
E30: PR +$C$30+($D$28-$D$27)*2/4
F30: PR +$C$30+($D$28-$D$27)*3/4
G30: PR +$C$30+($D$28-$D$27)*4/4
H30: PR [W2] '+
B31: PR [W12] 'Time (years)
C31: (F2) PR +C30/52
D31: (F2) PR +D30/52
E31: (F2) PR +E30/52
F31: (F2) PR +F30/52
G31: (F2) PR +G30/52
H31: PR [W2] '+
A32: PR [W17] 'Radial Time factor Tr
C32: (F2) PR +$D$9*C31/$E$21^2
D32: (F2) PR +$D$9*D31/$E$21^2
E32: (F2) PR +$D$9*E31/$E$21^2
F32: (F2) PR +$D$9*F31/$E$21^2
G32: (F2) PR +$D$9*G31/$E$21^2
H32: PR [W2] '+
A33: PR [W17] 'Deg of radial consol Ur (%)
C33: (F1) PR (1-@EXP((-8)*C32/$E$23))*100
D33: (F1) PR (1-@EXP((-8)*D32/$E$23))*100
E33: (F1) PR (1-@EXP((-8)*E32/$E$23))*100
F33: (F1) PR (1-@EXP((-8)*F32/$E$23))*100
G33: (F1) PR (1-@EXP((-8)*G32/$E$23))*100
H33: PR [W2] '+
A34: PR [W17] 'Vertical Time factor Tv
C34: (F2) PR +$D$8*C31/($D$10/2)^2
D34: (F2) PR +$D$8*D31/($D$10/2)^2
E34: (F2) PR +$D$8*E31/($D$10/2)^2
F34: (F2) PR +$D$8*F31/($D$10/2)^2

G34: (F2) PR +$D$8*G31/($D$10/2)^2
H34: PR [W2] '+
A35: PR [W17] 'Deg of vert consol Uv (%)
C35: (F1) PR @IF(C34<0.283,100*@SQRT(4*C34/@PI),100-10^((1.781-C34)/0.933))
D35: (F1) PR @IF(D34<0.283,100*@SQRT(4*D34/@PI),100-10^((1.781-D34)/0.933))
E35: (F1) PR @IF(E34<0.283,100*@SQRT(4*E34/@PI),100-10^((1.781-E34)/0.933))
F35: (F1) PR @IF(F34<0.283,100*@SQRT(4*F34/@PI),100-10^((1.781-F34)/0.933))
G35: (F1) PR @IF(G34<0.283,100*@SQRT(4*G34/@PI),100-10^((1.781-G34)/0.933))
H35: PR [W2] '+
A36: PR [W17] 'Deg of Consol (rad+vert)%
C36: (F1) PR (1-(1-C33/100)*(1-C35/100))*100
D36: (F1) PR (1-(1-D33/100)*(1-D35/100))*100
E36: (F1) PR (1-(1-E33/100)*(1-E35/100))*100
F36: (F1) PR (1-(1-F33/100)*(1-F35/100))*100
G36: (F1) PR (1-(1-G33/100)*(1-G35/100))*100
H36: PR [W2] '+
A37: PR [W17] \+
B37: PR [W12] \+
C37: PR \+
D37: PR \+
E37: PR \+
F37: PR \+
G37: PR \+
H37: PR [W2] '+
A38: PR [W17] 'Height of
D38: PR 'Settlements (mm) expected
H38: PR [W2] '+
A39: PR [W17] 'Fill (metres)
B39: PR [W12] "Total
D39: PR 'with   v e r t i c a l   d r a i n s
H39: PR [W2] '+
A40: PR [W17] *
B40: PR [W12] *
C40: PR *
D40: PR *
E40: PR *
F40: PR *
G40: PR *
H40: PR [W2] '+
A41: (F1) PR [W17] +D12
B41: (F1) PR [W12] +A41*$D$11*$D$10*$D$7*1000
C41: (F1) PR +B41*C36/100
D41: (F1) PR +B41*$D$36/100
E41: (F1) PR +B41*$E$36/100
F41: (F1) PR +B41*$F$36/100
G41: (F1) PR +B41*$G$36/100
H41: PR [W2] '+
A42: (F1) PR [W17] +A41+$D$13
B42: (F1) PR [W12] +A42*$D$11*$D$10*$D$7*1000
C42: (F1) PR +B42*$C$36/100
D42: (F1) PR +B42*$D$36/100
E42: (F1) PR +B42*$E$36/100
F42: (F1) PR +B42*$F$36/100
G42: (F1) PR +B42*$G$36/100
H42: PR [W2] '+
A43: (F1) PR [W17] +A42+$D$13
B43: (F1) PR [W12] +A43*$D$11*$D$10*$D$7*1000

```
C43: (F1) PR +B43*$C$36/100
D43: (F1) PR +B43*$D$36/100
E43: (F1) PR +B43*$E$36/100
F43: (F1) PR +B43*$F$36/100
G43: (F1) PR +B43*$G$36/100
H43: PR [W2] '+
A44: (F1) PR [W17] +A43+$D$13
B44: (F1) PR [W12] +A44*$D$11*$D$10*$D$7*1000
C44: (F1) PR +B44*$C$36/100
D44: (F1) PR +B44*$D$36/100
E44: (F1) PR +B44*$E$36/100
F44: (F1) PR +B44*$F$36/100
G44: (F1) PR +B44*$G$36/100
H44: PR [W2] '+
A45: (F1) PR [W17] +A44+$D$13
B45: (F1) PR [W12] +A45*$D$11*$D$10*$D$7*1000
C45: (F1) PR +B45*$C$36/100
D45: (F1) PR +B45*$D$36/100
E45: (F1) PR +B45*$E$36/100
F45: (F1) PR +B45*$F$36/100
G45: (F1) PR +B45*$G$36/100
H45: PR [W2] '+
A46: (F1) PR [W17] +A45+$D$13
B46: (F1) PR [W12] +A46*$D$11*$D$10*$D$7*1000
C46: (F1) PR +B46*$C$36/100
D46: (F1) PR +B46*$D$36/100
E46: (F1) PR +B46*$E$36/100
F46: (F1) PR +B46*$F$36/100
G46: (F1) PR +B46*$G$36/100
H46: PR [W2] '+
A47: PR [W17] \*
B47: PR [W12] \*
C47: PR \*
D47: PR \*
E47: PR \*
F47: PR \*
G47: PR \*
H47: PR [W2] '+
A49: PR [W17] \\
B49: PR [W12] \\
C49: PR \\
D49: PR \\
E49: PR \\
F49: PR \\
G49: PR \\
H49: PR [W2] '\
A50: PR [W17] 'Graph Aids to help illustrate the results
H50: PR [W2] '\
A51: PR [W17] 'Design Settlement line
E51: PR 'for fill height of
F51: PR +D12
G51: PR 'metres
H51: PR [W2] '\
H52: PR [W2] '\
A53: PR [W17] 'Text illustrating line
C53: (F0) PR +$B$46
D53: (F0) PR +$B$46
```

E53: (F0) PR +$B$46
F53: (F0) PR +$B$46
G53: (F0) PR +$B$46
H53: PR [W2] '\
C54: PR +C53-25
D54: PR +D53-25
E54: PR +E53-25
F54: PR +F53-25
G54: PR +G53-25
H54: PR [W2] '\
H55: PR [W2] '\
C56: PR 'Drain Radius (mm)
E56: PR @IF(D18>1000,"No vertical Drains",+D17*1000)
H56: PR [W2] '\
C57: PR 'Drain Spacing (m)
E57: PR @IF(+D18>100,"No vertical Drains",D18)
H57: PR [W2] '\
C58: PR @IF(+D19="t","Triangular pattern","Square pattern")
H58: PR [W2] '\
H59: PR [W2] '\
A60: PR [W17] \\
B60: PR [W12] \\
C60: PR \\
D60: PR \\
E60: PR \\
F60: PR \\
G60: PR \\
H60: PR [W2] '\

13. FILE: EMBANK2

A1: PR [W13] 'EMBANKMANT ANALYSIS AND GROUND IMPROVEMENT TECHNIQUES
A2: PR [W13] \-
B2: PR [W14] \-
C2: PR \-
D2: PR \-
E2: PR \-
B3: PR [W14] 'JOB DESCRIPTION :
D3: U 'Coomera
C5: PR 'EMBANKMENT AND SURCHARGE PROPERTIES
I5: PR ^NOTES
A6: PR [W13] \+
B6: PR [W14] \+
C6: PR \+
D6: PR \+
E6: PR \+
F6: PR \+
G6: PR [W2] \+
H6: PR \+
I6: PR \+
J6: PR \+
K6: PR \+
L6: PR \+
M6: PR '+
A7: PR [W13] 'Design Height of Embankment (H)

D7: U 2.5
E7: PR 'metres
G7: PR [W2] '+
M7: PR '+
A8: PR [W13] 'Embankment Width at top
D8: U 5
E8: PR 'metres
G8: PR [W2] '+
H8: PR 'Full width = 2 * B1
M8: PR '+
A9: PR [W13] 'Slope of Embankment is 1 Vertical :
D9: U 2
E9: PR 'Horizontal
G9: PR [W2] '+
H9: PR ' ==> Slope width (B2) =
K9: PR +D7*D9
L9: PR 'metres
M9: PR '+
A10: PR [W13] 'Analysis Increment surcharge Height
D10: U 1
E10: PR 'metres
G10: PR [W2] '+
H10: PR '{alpha1} =
I10: (F4) PR @ATAN(($D$8/2+$K$9)/$D$28)-@ATAN($D$8/2/$D$28)
J10: PR 'radians
M10: PR '+
A11: PR [W13] 'Unit weight of placed fill
D11: U 18
E11: PR 'kN / cu metre
G11: PR [W2] '+
H11: PR '{alpha2} =
I11: (F4) PR @ATAN($D$8/2/$D$28)
J11: PR 'radians
M11: PR '+
G12: PR [W2] '+
M12: PR '+
D13: PR 'VERTICAL DRAIN DESIGN
G13: PR [W2] '+
M13: PR '+
A14: PR [W13] \+
B14: PR [W14] \+
C14: PR \+
D14: PR \+
E14: PR \+
F14: PR \+
G14: PR [W2] \+
H14: PR \+
I14: PR \+
J14: PR \+
K14: PR \+
L14: PR \+
M14: PR '+
A15: PR [W13] '(T)riangular or (S)quare pattern
D15: U ^s
G15: PR [W2] '+
M15: PR '+
B16: PR [W14] "(W)ick or (S)and Drains

D16: U ^s
G16: PR [W2] '+
M16: PR '+
B17: PR [W14] 'Spacing of Drains S
D17: U 3
E17: PR 'metres
G17: PR [W2] '+
H17: PR '(Use S = 100 for no vertical drains)
M17: PR '+
B18: PR [W14] @IF(D16="w","Width of wick drain, b","Radius of sand drain,rd")
D18: U 0.15
E18: PR 'metres
G18: PR [W2] '+
H18: PR 'Equivalent Spacing de
K18: PR @IF(D15="s",0.564*D17,0.525*D17)
L18: PR 'metres
M18: PR '+
B19: PR [W14] 'Smear Factor
D19: U 1.5
E19: PR '* de
G19: PR [W2] '+
H19: PR 'Equivalent Drain Radius rw
K19: PR @IF(D16="w",(D18+D19)/2,D18)
L19: PR 'metres
M19: PR '+
G20: PR [W2] '+
H20: PR 'Factor de for smear effect
K20: PR +K18*D19
L20: PR 'metres
M20: PR '+
D21: PR 'TIME FRAMES
G21: PR [W2] '+
H21: PR "n =
I21: PR +K20/2/K19
K21: PR "m =
L21: PR +I21^2/(I21^2-1)*@LN(I21)-(3*I21^2-1)/4/I21^2
M21: PR '+
A22: PR [W13] \+
B22: PR [W14] \+
C22: PR \+
D22: PR \+
E22: PR \+
F22: PR \+
G22: PR [W2] \+
H22: PR \+
I22: PR \+
J22: PR \+
K22: PR \+
L22: PR \+
M22: PR '+
B23: PR [W14] 'Start of time t0
D23: U 0
E23: PR 'weeks
G23: PR [W2] '+
M23: PR '+
B24: PR [W14] 'End of time t1
D24: U 40

E24: PR 'weeks
G24: PR [W2] '+
H24: PR 'Time constraint for project
M24: PR '+
G25: PR [W2] '+
M25: PR '+
D26: PR 'SOIL PROFILE
G26: PR [W2] '+
M26: PR '+
A27: PR [W13] \+
B27: PR [W14] \+
C27: PR \+
D27: PR \+
E27: PR \+
F27: PR \+
G27: PR [W2] \+
H27: PR \+
I27: PR \+
J27: PR \+
K27: PR \+
L27: PR \+
M27: PR '+
A28: PR [W13] 'Depth to top of clay layer (z)
D28: U 0.1
E28: PR 'metres
G28: PR [W2] '+
H28: PR 'Near surface z = .01
M28: PR '+
A29: PR [W13] 'Unit weight of overburden soil
D29: U 17
E29: PR 'kN/ cu m
G29: PR [W2] '+
M29: PR '+
A30: PR [W13] 'Thickness of compressible layer h
D30: U 5
E30: PR 'metres
G30: PR [W2] '+
M30: PR '+
A31: PR [W13] 'Unit weight of clay layer
D31: U 15
E31: PR 'kN/ cu m
G31: PR [W2] '+
M31: PR '+
A32: PR [W13] 'Level of Water table below surface
D32: U 3
E32: PR ^metres
G32: PR [W2] '+
H32: PR 'Overburden Pressure (Po) at middle of
M32: PR '+
G33: PR [W2] '+
H33: PR 'compressible layer =
J33: PR @IF(D28<D32,@IF(D28+D30/2<D32,+D28*D29+(D31*D30/2),D28*D29+D31*
(D32-D28)+ (D31-9.81)*(D28+D30/2-D32)),(D29*D32)+(D29-9.81)*(D28-D32)+(D31-9.81)*D30/2)
K33: PR 'kPa
M33: PR '+
B34: PR [W14] 'SOIL PROPERTIES OF COMPRESSIBLE LAYER
G34: PR [W2] '+

M34: PR '+
A35: PR [W13] \+
B35: PR [W14] \+
C35: PR \+
D35: PR \+
E35: PR \+
F35: PR \+
G35: PR [W2] \+
H35: PR \+
I35: PR \+
J35: PR \+
K35: PR \+
L35: PR \+
M35: PR '+
A36: PR [W13] '   Preconsolidation Pressure Pc
D36: U 25
E36: PR ^kPa
G36: PR [W2] '+
H36: PR @IF(D36<=J33," Normally Consolidated Soil","Over consolidated soil")
M36: PR '+
B37: PR [W14] 'Compression Index Cc
D37: U 0.75
G37: PR [W2] '+
M37: PR '+
B38: PR [W14] 'Swelling Index Cs
G38: PR [W2] '+
H38: PR @IF(D36<=J33,"Cs value is not req'd for Norm. Consol. Soil","Cs is usually 5 to 10% of
Cc value")
M38: PR '+
A39: PR [W13] 'Initial Void Ratio eo
D39: U 1.6
G39: PR [W2] '+
M39: PR '+
A40: PR [W13] 'Coeff. of Consol.(Vertical) Cv
D40: U 5
E40: PR 'sq m / yr
G40: PR [W2] '+
M40: PR '+
A41: PR [W13] 'Coeff. of Consol. (Radial) Cvr
D41: U 10
E41: PR 'sq m / yr
G41: PR [W2] '+
M41: PR '+
A42: PR [W13] 'Undrained Cohesion
D42: U 50
E42: PR 'kPa
G42: PR [W2] '+
H42: PR '   Allowable Stress increase =
K42: (F2) PR (2+@PI)*D42/D43
L42: PR 'kPa
M42: PR '+
A43: PR [W13] 'Bearing Capacity safety factor
D43: U 2.5
G43: PR [W2] '+
M43: PR '+
G44: PR [W2] '+
M44: PR '+

C45: PR 'DESIGN CONSIDERATIONS
G45: PR [W2] '+
M45: PR '+
A46: PR [W13] \+
B46: PR [W14] \+
C46: PR \+
D46: PR \+
E46: PR \+
F46: PR \+
G46: PR [W2] \+
H46: PR \+
I46: PR \+
J46: PR \+
K46: PR \+
L46: PR \+
M46: PR '+
A47: PR [W13] 'The following should be noted :
G47: PR [W2] '+
M47: PR '+
A48: PR [W13] '- A Maximum Embankment Height (H) of
D48: (F3) PR +K42/D11/K56
E48: PR 'metres is required
G48: PR [W2] '+
M48: PR '+
A49: PR [W13] ' to avoid Bearing capacity failure
G49: PR [W2] '+
M49: PR '+
A50: PR [W13] @IF($D$7<$D$48,"- Embankment may be constructed to full Final Height","- A staged loading will be required to achieve Final Height")
G50: PR [W2] '+
M50: PR '+
A51: PR [W13] '- A constant Side slope and embankment width assumed, and is
G51: PR [W2] '+
M51: PR '+
A52: PR [W13] ' independent of surcharge height
G52: PR [W2] '+
M52: PR '+
A53: PR [W13] '- The Slope stability of the embankment should be checked
G53: PR [W2] '+
M53: PR '+
A54: PR [W13] '- Fill Height for settlemment analysis will be constrained by
G54: PR [W2] '+
M54: PR '+
A55: PR [W13] ' * The maximum height which will not cause Bearing failure
G55: PR [W2] '+
M55: PR '+
A56: PR [W13] ' * Side slopes (1V : H) and embankment width (2* B1) specified
G56: PR [W2] '+
H56: PR 'Influence Factor (I) of
K56: (F4) PR (($D$8/2+$K$9)/$K$9*($I$10+$I$11)-$D$8/2/$K$9*$I$11)/@PI*2
L56: PR ^below
M56: PR '+
A57: PR [W13] '- Influence factor is calculated to middle of the compressible
G57: PR [W2] '+
H57: PR 'middle of embankment & TOP of compress. layer
M57: PR '+
A58: PR [W13] ' layer & middle of the embankment in the analysis below

G58: PR [W2] '+
M58: PR '+
A59: PR [W13] '- Average increase in pressure over the layer is used
G59: PR [W2] '+
M59: PR '+
G60: PR [W2] '+
M60: PR '+
G61: PR [W2] '+
H61: PR 'CALCULATION OF CONSOLIDATION PARAMATERS
M61: PR '+
A62: PR [W13] *
B62: PR [W14] *
C62: PR *
D62: PR *
E62: PR *
F62: PR *
G62: PR [W2] *
H62: PR *
I62: PR *
J62: PR *
K62: PR *
L62: PR *
M62: PR '*
E63: PR 'Time (weeks)
G63: PR [W2] ":
H63: PR +D23
I63: PR +$H$63+($D$24-$D$23)/4
J63: PR +$H$63+($D$24-$D$23)*2/4
K63: PR +$H$63+($D$24-$D$23)*3/4
L63: PR +$H$63+($D$24-$D$23)*4/4
M63: PR '*
E64: PR 'Time (years)
G64: PR [W2] ":
H64: (F2) PR +I163/52
I64: (F2) PR +I63/52
J64: (F2) PR +J63/52
K64: (F2) PR +K63/52
L64: (F2) PR +L63/52
M64: PR '*
E65: PR 'Radial Time factor Tr
G65: PR [W2] ":
H65: (F2) PR +$D$41*H64/$K$20^2
I65: (F2) PR +$D$41*I64/$K$20^2
J65: (F2) PR +$D$41*J64/$K$20^2
K65: (F2) PR +$D$41*K64/$K$20^2
L65: (F2) PR +$D$41*L64/$K$20^2
M65: PR '*
C66: PR 'Deg of radial consolidation Ur (%)
G66: PR [W2] ":
H66: (F1) PR (1-@EXP((-8)*H65/$L$21))*100
I66: (F1) PR (1-@EXP((-8)*I65/$L$21))*100
J66: (F1) PR (1-@EXP((-8)*J65/$L$21))*100
K66: (F1) PR (1-@EXP((-8)*K65/$L$21))*100
L66: (F1) PR (1-@EXP((-8)*L65/$L$21))*100
M66: PR '*
D67: PR 'Vertical Time factor Tv
G67: PR [W2] ":

H67: (F2) PR +$D$40*H64/($D$30/2)^2
I67: (F2) PR +$D$40*I64/($D$30/2)^2
J67: (F2) PR +$D$40*J64/($D$30/2)^2
K67: (F2) PR +$D$40*K64/($D$30/2)^2
L67: (F2) PR +$D$40*L64/($D$30/2)^2
M67: PR '*
C68: PR 'Deg of vertical consolidation Uv (%)
G68: PR [W2] ":
H68: (F1) PR @IF(H67<0.283,100*@SQRT(4*H67/@PI),100-10^((1.781-H67)/0.933))
I68: (F1) PR @IF(I67<0.283,100*@SQRT(4*I67/@PI),100-10^((1.781-I67)/0.933))
J68: (F1) PR @IF(J67<0.283,100*@SQRT(4*J67/@PI),100-10^((1.781-J67)/0.933))
K68: (F1) PR @IF(K67<0.283,100*@SQRT(4*K67/@PI),100-10^((1.781-K67)/0.933))
L68: (F1) PR @IF(L67<0.283,100*@SQRT(4*L67/@PI),100-10^((1.781-L67)/0.933))
M68: PR '*
B69: PR [W14] 'Degree of Consolidation (radial + vertical) %
G69: PR [W2] ":
H69: (F1) PR (1-(1-H66/100)*(1-H68/100))*100
I69: (F1) PR (1-(1-I66/100)*(1-I68/100))*100
J69: (F1) PR (1-(1-J66/100)*(1-J68/100))*100
K69: (F1) PR (1-(1-K66/100)*(1-K68/100))*100
L69: (F1) PR (1-(1-L66/100)*(1-L68/100))*100
M69: PR '*
A70: PR [W13] *
B70: PR [W14] *
C70: PR *
D70: PR *
E70: PR *
F70: PR *
G70: PR [W2] *
H70: PR *
I70: PR *
J70: PR *
K70: PR *
L70: PR *
M70: PR '*
M71: PR '*
F72: PR ^CALCULATION OF INFLUENCE FACTORS WITH DEPTH
M72: PR '*
A73: PR [W13] *
B73: PR [W14] *
C73: PR *
D73: PR *
E73: PR *
F73: PR *
G73: PR [W2] *
H73: PR *
I73: PR *
J73: PR *
K73: PR *
L73: PR *
M73: PR '*
B74: PR [W14] 'Height of
C74: PR ^Slope
D74: PR '   { a l p h a 1 }
I74: PR '   INFLUENCE FACTOR (I)
L74: PR ^Avg (kPa)
M74: PR '*

```
B75: PR [W14] 'Fill (metres)
C75: PR ^Width
D75: PR ^Middle
E75: PR ^Top
F75: PR ^Bottom
I75: PR ^Middle
J75: PR ^Top
K75: PR ^Bottom
L75: PR ^Pressure
M75: PR '*
B76: PR [W14] ^h
C76: PR ^(B2)
H76: PR '{alpha2}=
I76: (F4) PR @ATAN($D$8/2/($D$28+$D$30/2))
J76: (F4) PR @ATAN($D$8/2/($D$28))
K76: (F4) PR @ATAN($D$8/2/($D$28+$D$30))
L76: PR ^Increase
M76: PR '*
A77: PR [W13] \*
B77: PR [W14] \*
C77: PR \*
D77: PR \*
E77: PR \*
F77: PR \*
G77: PR [W2] \*
H77: PR \*
I77: PR \*
J77: PR \*
K77: PR \*
L77: PR \*
M77: PR '*
B78: (F1) PR [W14] +$D$7
C78: (F2) PR +B78*$D$9
D78: (F4) PR @ATAN(($D$8/2+$C$78)/($D$28+$D$30/2))-@ATAN($D$8/2/($D$28+$D$30/2))
E78: (F4) PR @ATAN(($D$8/2+C78)/($D$28))-@ATAN($D$8/2/($D$28))
F78: (F4) PR @ATAN(($D$8/2+C78)/($D$28+$D$30))-@ATAN($D$8/2/($D$28+$D$30))
I78: (F3) PR (($D$8/2+$C$78)/$C$78*($D$78+$I$76)-$D$8/2/$C$78*I76)/@PI*2
J78: (F3) PR (($D$8/2+$C$78)/$C$78*($E$78+$I$76)-$D$8/2/$C$78*J76)/@PI*2
K78: (F3) PR (($D$8/2+$C$78)/$C$78*($F$78+$I$76)-$D$8/2/$C$78*K76)/@PI*2
L78: (F1) PR (+J78+4*I78+K78)/6*B78*$D$11
M78: PR '*
B79: (F1) PR [W14] +B78+$D$10
C79: (F2) PR +B79*$D$9
D79: (F4) PR @ATAN(($D$8/2+C79)/($D$28+$D$30/2))-@ATAN($D$8/2/($D$28+$D$30/2))
E79: (F4) PR @ATAN(($D$8/2+C79)/($D$28))-@ATAN($D$8/2/($D$28))
F79: (F4) PR @ATAN(($D$8/2+C79)/($D$28+$D$30))-@ATAN($D$8/2/($D$28+$D$30))
I79: (F3) PR (($D$8/2+C79)/C79*(D79+$I$76)-$D$8/2/C79*$I$76)/@PI*2
J79: (F3) PR (($D$8/2+$C$78)/$C$78*($E$78+$I$76)-$D$8/2/$C$78*J74)/@PI*2
K79: (F3) PR (($D$8/2+$C$78)/$C$78*($F$78+$I$76)-$D$8/2/$C$78*K74)/@PI*2
L79: (F1) PR (+J79+4*I79+K79)/6*B79*$D$11
M79: PR '*
B80: (F1) PR [W14] +B79+$D$10
C80: (F2) PR +B80*$D$9
D80: (F4) PR @ATAN(($D$8/2+C80)/($D$28+$D$30/2))-@ATAN($D$8/2/($D$28+$D$30/2))
E80: (F4) PR @ATAN(($D$8/2+C80)/($D$28))-@ATAN($D$8/2/($D$28))
F80: (F4) PR @ATAN(($D$8/2+C80)/($D$28+$D$30))-@ATAN($D$8/2/($D$28+$D$30))
I80: (F3) PR (($D$8/2+C80)/C80*(D80+$I$76)-$D$8/2/C80*$I$76)/@PI*2
```

J80: (F3) PR (($D$8/2+$C$78)/$C$78*($E$78+$I$76)-$D$8/2/$C$78*J75)/@PI*2
K80: (F3) PR (($D$8/2+$C$78)/$C$78*($F$78+$I$76)-$D$8/2/$C$78*K75)/@PI*2
L80: (F1) PR (+J80+4*I80+K80)/6*B80*$D$11
M80: PR '*
B81: (F1) PR [W14] +B80+$D$10
C81: (F2) PR +B81*$D$9
D81: (F4) PR @ATAN(($D$8/2+C81)/($D$28+$D$30/2))-@ATAN($D$8/2/($D$28+$D$30/2))
E81: (F4) PR @ATAN(($D$8/2+C81)/($D$28))-@ATAN($D$8/2/($D$28))
F81: (F4) PR @ATAN(($D$8/2+C81)/($D$28+$D$30))-@ATAN($D$8/2/($D$28+$D$30))
I81: (F3) PR (($D$8/2+C81)/C81*(D81+$I$76)-$D$8/2/C81*$I$76)/@PI*2
J81: (F3) PR (($D$8/2+$C$78)/$C$78*($E$78+$I$76)-$D$8/2/$C$78*J77)/@PI*2
K81: (F3) PR (($D$8/2+$C$78)/$C$78*($F$78+$I$76)-$D$8/2/$C$78*K77)/@PI*2
L81: (F1) PR (+J81+4*I81+K81)/6*B81*$D$11
M81: PR '*
B82: (F1) PR [W14] +B81+$D$10
C82: (F2) PR +B82*$D$9
D82: (F4) PR @ATAN(($D$8/2+C82)/($D$28+$D$30/2))-@ATAN($D$8/2/($D$28+$D$30/2))
E82: (F4) PR @ATAN(($D$8/2+C82)/($D$28))-@ATAN($D$8/2/($D$28))
F82: (F4) PR @ATAN(($D$8/2+C82)/($D$28+$D$30))-@ATAN($D$8/2/($D$28+$D$30))
I82: (F3) PR (($D$8/2+C82)/C82*(D82+$I$76)-$D$8/2/C82*$I$76)/@PI*2
J82: (F3) PR (($D$8/2+$C$78)/$C$78*($E$78+$I$76)-$D$8/2/$C$78*J78)/@PI*2
K82: (F3) PR (($D$8/2+$C$78)/$C$78*($F$78+$I$76)-$D$8/2/$C$78*K78)/@PI*2
L82: (F1) PR (+J82+4*I82+K82)/6*B82*$D$11
M82: PR '*
B83: (F1) PR [W14] +B82+$D$10
C83: (F2) PR +B83*$D$9
D83: (F4) PR @ATAN(($D$8/2+C83)/($D$28+$D$30/2))-@ATAN($D$8/2/($D$28+$D$30/2))
E83: (F4) PR @ATAN(($D$8/2+C83)/($D$28))-@ATAN($D$8/2/($D$28))
F83: (F4) PR @ATAN(($D$8/2+C83)/($D$28+$D$30))-@ATAN($D$8/2/($D$28+$D$30))
I83: (F3) PR (($D$8/2+C83)/C83*(D83+$I$76)-$D$8/2/C83*$I$76)/@PI*2
J83: (F3) PR (($D$8/2+$C$78)/$C$78*($E$78+$I$76)-$D$8/2/$C$78*J79)/@PI*2
K83: (F3) PR (($D$8/2+$C$78)/$C$78*($F$78+$I$76)-$D$8/2/$C$78*K79)/@PI*2
L83: (F1) PR (+J83+4*I83+K83)/6*B83*$D$11
M83: PR '*
A84: PR [W13] *
B84: PR [W14] *
C84: PR *
D84: PR *
E84: PR *
F84: PR *
G84: PR [W2] *
H84: PR *
I84: PR *
J84: PR *
K84: PR *
L84: PR *
M84: PR '*
H89: PR 'CALCULATION OF SETTLEMENTS with TIME
C90: PR "*
D90: PR *
E90: PR *
F90: PR *
G90: PR [W2] *
H90: PR *
I90: PR *
J90: PR *
K90: PR *

L90: PR *
M90: PR '*
C91: PR "*
G91: PR [W2] '|
I91: PR ^T I M E (weeks)
M91: PR '*
C92: PR "*
D92: PR 'Height
E92: PR ^Average
F92: PR ^Total
G92: PR [W2] '|
H92: PR +H63
I92: PR +I63
J92: PR +J63
K92: PR +K63
L92: PR +L63
M92: PR '*
C93: PR "*
D93: PR ^of
E93: PR ^Pressure
F93: PR ^Induced
G93: PR [W2] '|
H93: PR \-
I93: PR \-
J93: PR \-
K93: PR \-
L93: PR \-
M93: PR '*
C94: PR "*
D94: PR ^Fill
E94: PR ^Increase
F94: PR "Settlement
G94: PR [W2] '|
M94: PR '*
C95: PR "*
G95: PR [W2] '|
I95: PR 'Settlements (mm) expected
L95: PR 'with time
M95: PR '*
C96: PR "*
D96: PR ^(metres)
E96: PR ^(kPa)
F96: PR ^(mm)
G96: PR [W2] '|
I96: PR @IF(D17<100,"and   v e r t i c a l   d r a i n s","No V e r t i c a l D r a i n s")
M96: PR '*
C97: PR "*
D97: PR *
E97: PR *
F97: PR *
G97: PR [W2] *
H97: PR *
I97: PR *
J97: PR *
K97: PR *
L97: PR *
M97: PR '*

```
C98: PR "*
D98: (F1) PR +B78
E98: (F1) PR +L78
F98: (F1) PR @IF($D$36<=$J$33,+$D$37*$D$30/(1+$D$39)*@LOG((+$J$33+E98)/$J$33)*
1000,@ IF(($J$33+E98)<$D$36,$D$38*$D$30/(1+$D$39)*@LOG(($J$33+E98)/$J$33)*1000,$D$38*
$D$30/(1+$D$39)*@LOG($D$36/$J$33)+$D$37*$D$30/(1+$D$39)*@LOG(($J$33+E98)/$D$36)
H98: (F0) PR +F98*H69/100
I98: (F0) PR +F98*$I$69/100
J98: (F0) PR +F98*$J$69/100
K98: (F0) PR +F98*$K$69/100
L98: (F0) PR +F98*$L$69/100
M98: PR '*
C99: PR "*
D99: (F1) PR +B79
E99: (F1) PR +L79
F99: (F1) PR @IF($D$36<=$J$33,+$D$37*$D$30/(1+$D$39)*@LOG((+$J$33+E99)/$J$33)*
1000,@ IF(($J$33+E99)<$D$36,$D$38*$D$30/(1+$D$39)*@LOG(($J$33+E99)/$J$33)*1000,$D$38*
$D$30/(1+$D$39)*@LOG($D$36/$J$33)+$D$37*$D$30/(1+$D$39)*@LOG(($J$33+E99)/$D$36)
H99: (F0) PR +F99*$H$69/100
I99: (F0) PR +F99*$I$69/100
J99: (F0) PR +F99*$J$69/100
K99: (F0) PR +F99*$K$69/100
L99: (F0) PR +F99*$L$69/100
M99: PR '*
C100: PR "*
D100: (F1) PR +B80
E100: (F1) PR +L80
F100: (F1) PR @IF($D$36<=$J$33,+$D$37*$D$30/(1+$D$39)*@LOG((+$J$33+E100)/$J$33)*
1000, @IF(($J$33+E100)<$D$36,$D$38*$D$30/(1+$D$39)*@LOG(($J$33+E100)/$J$33)*1000,$D$38
*$D$30/(1+$D$39)*@LOG($D$36/$J$33)+$D$37*$D$30/(1+$D$39)*@LOG(($J$33+E100)/$D$36)))
H100: (F0) PR +F100*$H$69/100
I100: (F0) PR +F100*$I$69/100
J100: (F0) PR +F100*$J$69/100
K100: (F0) PR +F100*$K$69/100
L100: (F0) PR +F100*$L$69/100
M100: PR '*
C101: PR "*
D101: (F1) PR +B81
E101: (F1) PR +L81
F101: (F1) PR @IF($D$36<=$J$33,+$D$37*$D$30/(1+$D$39)*@LOG((+$J$33+E101)/$J$33)*
1000, @IF(($J$33+E101)<$D$36,$D$38*$D$30/(1+$D$39)*@LOG(($J$33+E101)/$J$33)*1000,$D$38
*$D$30/(1+$D$39)*@LOG($D$36/$J$33)+$D$37*$D$30/(1+$D$39)*@LOG(($J$33+E101)/$D$36)))
H101: (F0) PR +F101*$H$69/100
I101: (F0) PR +F101*$I$69/100
J101: (F0) PR +F101*$J$69/100
K101: (F0) PR +F101*$K$69/100
L101: (F0) PR +F101*$L$69/100
M101: PR '*
C102: PR "*
D102: (F1) PR +B82
E102: (F1) PR +L82
F102: (F1) PR @IF($D$36<=$J$33,+$D$37*$D$30/(1+$D$39)*@LOG((+$J$33+E102)/$J$33)*
1000, @IF(($J$33+E102)<$D$36,$D$38*$D$30/(1+$D$39)*@LOG(($J$33+E102)/$J$33)*1000,$D$38
*$D$30/(1+$D$39)*@LOG($D$36/$J$33)+$D$37*$D$30/(1+$D$39)*@LOG(($J$33+E102)/$D$36)))
H102: (F0) PR +F102*$H$69/100
I102: (F0) PR +F102*$I$69/100
J102: (F0) PR +F102*$J$69/100
```

```
K102: (F0) PR +F102*$K$69/100
L102: (F0) PR +F102*$L$69/100
M102: PR '*
C103: PR "*
D103: (F1) PR +B83
E103: (F1) PR +L83
F103: (F1) PR @IF($D$36<=$J$33,+$D$37*$D$30/(1+$D$39)*@LOG((+$J$33+E103)/$J$33)*
1000, @IF(($J$33+E103)<$D$36,$D$38*$D$30/(1+$D$39)*@LOG(($J$33+E103)/$J$33)*1000,$D$38
*$D$30/(1+$D$39)*@LOG($D$36/$J$33)+$D$37*$D$30/(1+$D$39)*@LOG(($J$33+E103)/$D$36)))
H103: (F0) PR +F103*$H$69/100
I103: (F0) PR +F103*$I$69/100
J103: (F0) PR +F103*$J$69/100
K103: (F0) PR +F103*$K$69/100
L103: (F0) PR +F103*$L$69/100
M103: PR '*
C104: PR "*
D104: PR \*
E104: PR \*
F104: PR \*
G104: PR [W2] \*
H104: PR \*
I104: PR \*
J104: PR \*
K104: PR \*
L104: PR \*
M104: PR '*
G107: PR [W2] 'GRAPH AIDS to illustrate results
E108: PR "\
F108: PR \\
G108: PR [W2] \\
H108: PR \\
I108: PR \\
J108: PR \\
K108: PR \\
L108: PR \\
M108: PR '\
E109: PR "\
F109: PR 'Design Settlement line
J109: PR 'for fill height of
K109: (F1) PR +B78
L109: PR 'metres
M109: PR '\
E110: PR "\
M110: PR '\
E111: PR "\
F111: PR '" Dummy "
H111: (F0) PR +$F$103-($F$103-$F$102)/2
I111: (F0) PR +$F$103-($F$103-$F$102)/2
J111: (F0) PR +$F$103-($F$103-$F$102)/2
K111: (F0) PR +$F$103-($F$103-$F$102)/2
L111: (F0) PR +$F$103-($F$103-$F$102)/2
M111: PR '\
E112: PR "\
F112: PR ^Lines
H112: (F0) PR +$F$102
I112: (F0) PR +$F$102
J112: (F0) PR +$F$102
```

K112: (F0) PR +$F$102
L112: (F0) PR +$F$102
M112: PR '\
E113: PR "\
M113: PR '\
E114: PR "\
H114: PR @IF(D16="w","Wick Width, mm","Drain Radius (mm)")
J114: PR @IF(D17>100,"No vertical Drains",+D18*1000)
M114: PR '\
E115: PR "\
H115: PR 'Drain Spacing, m
J115: PR @IF(+D17<100,D17,"No vertical Drains")
M115: PR '\
E116: PR "\
H116: PR @IF(+D15="t","Triangular pattern","Square pattern")
M116: PR '\
E117: PR "\
M117: PR '\
E118: PR "\
F118: PR \\
G118: PR [W2] \\
H118: PR \\
I118: PR \\
J118: PR \\
K118: PR \\
L118: PR \\
M118: PR '\

14. FILE: MACSTRP1

C1: 'STRESS VARIATION DUE TO POINT LOAD
A2: [W9] _
B2: [W15] _
C2: _
D2: _
E2: [W8] _
F2: [W9] _
B3: [W15] '   Surface Point Load (Q) = ?
E3: U [W8] 100
F3: [W9] '<- kN
B4: [W15] '   Analysis begins at a depth = ?
E4: U [W8] 10
F4: [W9] '<- metres
C5: '   Final Depth (z) = ?
E5: U [W8] 20
F5: [W9] '<- metres
C6: 'No of increments (N) = ?
E6: U [W8] 1
F6: [W9] '<-
B7: [W15] 'x - Distance from centre of loading = ?
E7: U [W8] 1
F7: [W9] '<- metres
B8: [W15] 'y - Distance from centre of loading = ?
E8: U [W8] 1
F8: [W9] '<- metres

A10: [W9] _
B10: [W15] _
C10: _
D10: _
E10: [W8] _
F10: [W9] _
A31: [W9] 'The highlighted cells are the data entry cells. Enter data
A32: [W9] 'in the highlighted cells (Cells E3 to E8). The worksheet
A33: [W9] 'calculates the Vertical & Shear stress variation with depth.
A35: [W9] 'FILL in the above data and press ALT + A to activate macro
A36: [W9] 'The macro will take a few seconds to execute, during which
A37: (F3) [W9] 'CMD will appear at the bottom centre of the screen.

AA1: '{paneloff}{windowsoff}
AA2: '/rncloop~ag21~
AA3: '/rncend~ag24~
AA4: '/rnccont~aa21~
AA5: '/rnccarryon~aa23~
AA6: '{goto}be9~
AA7: '@sqrt(e7^2+e8^2)~/rff3~~/rv~e9~
AA8: '{goto}f9~metres~{left 4}'=====>~
AA9: '{right}'r - Distance = ~
AA10: '{down}{left 2}\+~/c~{right}.{right 4}~
AA11: '{down}^Depth~{right}^Least Distance~
AA12: '{right}^VERTICAL~{right}^SHEAR~
AA13: '{down}^STRESS {{}kPa{}}~{left}^STRESS {{}kPa{}}~
AA14: '{left}^R {{}m{}}~{left}^metres~
AA15: '{down}\+~/c~{right}.{right 4}~
AA16: '{goto}ba14~+e4~/rv~a14~
AA17: '{right}@sqrt(e9^2+a14^2)~/rv~b14~
AA18: '{right}(3*e3*(a14)^3)/(2*@pi*(b14)^5)~/rv~c14~
AA19: '{right}(3*e3*(a14)^2*e9)/(2*@pi*(b14)^5)~/rv~d14~
AA20: '{goto}ba14~
AA21: '{if @cellpointer("contents")>=+e5}{branch end}
AG21: '{down}+{up}+((e5-e4)/e6)~
AA22: '{branch loop}
AG22: '{branch cont}
AA23: '{end}{up}/rff3~.{end}{down}~/rv.{end}{down}~a14~
AA24: '{end}{down 2}{right}99~/c~{right}.{right}~
AG24: '{branch carryon}
AA25: '{end}{up}/c~{down}.{end}{down}{up}~{end}{down}/re~{end}{up}{end}{up}/rff3~.{end}
{down}~/rv.{end}{down}~b14~
AA26: '{right}/c~{down}.{end}{down}{up}~{end}{down}/re~{end}{up}{end}{up}/rff2~.{end}
{down}~/rv.{end}{down}~c14~
AA27: '{right}/c~{down}.{end}{down}{up}~{end}{down}/re~{end}{up}{end}{up}/rff2~.{end}
{down}~/rv.{end}{down}~d14~
AA28: '{Home}{down 20}{beep}

15. FILE: MACSTRP2

A1: PR 'STRESS VARIATION DUE TO POINT LOAD
A2: PR "--------------
B2: PR [W14] "--------------
C2: PR [W13] "--------------

D2: PR "--------------
B4: PR [W14] ' Surface Point Load (Q) = ?
E4: U 100
F4: PR 'kN
B5: PR [W14] 'Analysis begins at a depth = ?
E5: U 1
F5: PR 'metres
B6: PR [W14] '   Increment depth (z) = ?
E6: U 0.5
F6: PR 'metres
A7: PR 'x - Distance from centre of loading = ?
E7: U 1
F7: PR 'metres
A8: PR 'y - Distance from centre of loading = ?
E8: U 1
F8: PR 'metres
B9: PR [W14] "=====>
C9: PR [W13] ' r - Distance =
E9: (F3) PR @SQRT(E7^2+E8^2)
F9: PR 'metres
A10: PR "+++++++++++++
B10: PR [W14] "+++++++++++++++
C10: PR [W13] "+++++++++++++
D10: PR "+++++++++++++
E10: PR "+++++++++++++
F10: PR "+++++++++++++
A11: PR 'Depth
B11: PR [W14] 'Least Distance
C11: PR [W13] ^VERTICAL
D11: PR ^SHEAR
A12: PR '{m}
B12: PR [W14] ^R {m}
C12: PR [W13] 'STRESS {kPa}
D12: PR 'STRESS {kPa}
A13: PR "+++++++++++++
B13: PR [W14] "+++++++++++++++
C13: PR [W13] "+++++++++++++
D13: PR "+++++++++++++
E13: PR "+++++++++++++
F13: PR "+++++++++++++
A14: (F3) PR +E5
B14: (F3) PR [W14] @SQRT(+A14^2+$E$9^2)
C14: (F2) PR [W13] 3*$E$4*A14^3/(2*@PI*B14^5)
D14: (F2) PR 3*$E$4*$E$9*A14^2/(2*@PI*B14^5)
A15: (F3) PR +A14+$E$6
B15: (F3) PR [W14] @SQRT(+A15^2+$E$9^2)
C15: (F2) PR [W13] 3*$E$4*A15^3/(2*@PI*B15^5)
D15: (F2) PR 3*$E$4*$E$9*A15^2/(2*@PI*B15^5)
A16: (F3) PR +A15+$E$6
B16: (F3) PR [W14] @SQRT(+A16^2+$E$9^2)
C16: (F2) PR [W13] 3*$E$4*A16^3/(2*@PI*B16^5)
D16: (F2) PR 3*$E$4*$E$9*A16^2/(2*@PI*B16^5)
A17: (F3) PR +A16+$E$6
B17: (F3) PR [W14] @SQRT(+A17^2+$E$9^2)
C17: (F2) PR [W13] 3*$E$4*A17^3/(2*@PI*B17^5)
D17: (F2) PR 3*$E$4*$E$9*A17^2/(2*@PI*B17^5)
A18: (F3) PR +A17+$E$6

B18: (F3) PR [W14] @SQRT(+A18^2+$E$9^2)
C18: (F2) PR [W13] 3*$E$4*A18^3/(2*@PI*B18^5)
D18: (F2) PR 3*$E$4*$E$9*A18^2/(2*@PI*B18^5)
A19: (F3) PR +A18+$E$6
B19: (F3) PR [W14] @SQRT(+A19^2+$E$9^2)
C19: (F2) PR [W13] 3*$E$4*A19^3/(2*@PI*B19^5)
D19: (F2) PR 3*$E$4*$E$9*A19^2/(2*@PI*B19^5)
A20: (F3) PR +A19+$E$6
B20: (F3) PR [W14] @SQRT(+A20^2+$E$9^2)
C20: (F2) PR [W13] 3*$E$4*A20^3/(2*@PI*B20^5)
D20: (F2) PR 3*$E$4*$E$9*A20^2/(2*@PI*B20^5)
A21: (F3) PR +A20+$E$6
B21: (F3) PR [W14] @SQRT(+A21^2+$E$9^2)
C21: (F2) PR [W13] 3*$E$4*A21^3/(2*@PI*B21^5)
D21: (F2) PR 3*$E$4*$E$9*A21^2/(2*@PI*B21^5)
A22: (F3) PR +A21+$E$6
B22: (F3) PR [W14] @SQRT(+A22^2+$E$9^2)
C22: (F2) PR [W13] 3*$E$4*A22^3/(2*@PI*B22^5)
D22: (F2) PR 3*$E$4*$E$9*A22^2/(2*@PI*B22^5)
A23: (F3) PR +A22+$E$6
B23: (F3) PR [W14] @SQRT(+A23^2+$E$9^2)
C23: (F2) PR [W13] 3*$E$4*A23^3/(2*@PI*B23^5)
D23: (F2) PR 3*$E$4*$E$9*A23^2/(2*@PI*B23^5)
A24: (F3) PR +A23+$E$6
B24: (F3) PR [W14] @SQRT(+A24^2+$E$9^2)
C24: (F2) PR [W13] 3*$E$4*A24^3/(2*@PI*B24^5)
D24: (F2) PR 3*$E$4*$E$9*A24^2/(2*@PI*B24^5)
A30: PR 'PRESS {ALT + M} KEYS TOGETHER TO VIEW MENU
A32: PR 'The highlighted cells seen above are the data entry cells.
A33: PR 'You can enter values only in these highlighted cells. The
A34: PR 'worksheet then calculates the stresss variation with depth.

GH2313: (H) PR '\M
GI2313: (H) PR '/xmgi2316~
GI2314: (H) PR '{GOTO}E4~
GI2316: (H) PR 'ENTER VALUES
GJ2316: (H) PR 'PRINT WORKSHEET
GK2316: (H) PR 'VIEW GRAPH
GL2316: (H) PR 'EXIT
GI2317: (H) PR 'Enter data in highlighted cells
GJ2317: (H) PR 'Print the existing worksheet screen including input data and output table
GK2317: (H) PR 'View the stress variation with depth graph from the worsheeet table
GL2317: (H) PR 'End the point load stress worksheet and quit Lotus program.
GI2318: (H) PR '{HOME}~{RIGHT 4}~{DOWN 3}~
GJ2318: (H) PR "/PPRa1.f25~GQ
GK2318: (H) PR '{graph}~
GL2318: (H) PR '/qy~

References

Carter, M. 1983. *Geotechnical engineering handbook*. Pentech Press.

Craig, R.F. 1983. *Soil mechanics* (3rd ed.). Van Nostrand Reinhold.

Das, B.M. 1984. *Principles of foundation engineering*. Brooks/Cole Engineering Division.

Geoguide 1. 1982. *Guide to retaining wall design*. Geotechnical Control Office, Hong Kong.

Hoek, E. & J.W. Bray 1981. *Rock slope engineering* (revised 3rd ed.). Institute of Mining and Metallurgy, London.

Ingold, T.S. 1979. The effects of compaction on retaining walls. *Geotechnique* 29.

Lotus 123 manuals.

Meyerhof, G.G. 1970. Safety factors in soil mechanics. *Canadian Geotechnical Journal* 7(4).

Meyerhof, G.G. 1976. Bearing capacity and settlement of pile foundations. *Proceedings of the American Society of Civil Engineers* 102(GT3): Eleventh Terzaghi Lecture.

NAVFAC 1986. Foundation and earth structures. *Design Manual* 7.02.

Pearson Kirk, D. 1976. Lateral pressures exerted by compacted granular materials. *Australian Road Research Proceedings*, Vol. 8.

Randolph, M.F. 1989. *PISSAP: Pile spreadsheet analysis program*. University of Western Australia.

Sage, R. 1977. Pit slope manual – Mechanical support. CANMET report 77-3.

Semple, R.M. 1981. Partial coefficient design in geotechnics. *Ground engineering*, September.

Simons, N.E. & B.K. Menzies 1979. *A short course in foundation engineering*: Butterworth.

Skempton, A.W. 1959. Cast in-situ bored piles in London clay. *Geotechnique* 9.

Smith, I.M. 1988. Geotechnical Aspects of the use of computers in engineering: A personal view. *Proceedings Institute of Civil Engineers*, Part 1: 565-569.

Stamatopoulos, A.C. & P.C. Kotzias 1985. *Soil improvement by preloading*. Wiley.

Tomlinson, M.J. 1970. Some effects of pile driving on skin friction. *Proceedings of the conference on behaviour of piles, Institute of Civil Engineers, London*.

Tomlinson, M.J. 1977. *Pile design and construction practices* Viewpoint Publications.

Tomlinson, M.J. 1980. *Foundation design and construction* (4th ed.). Pitman.

Subject index

SPREADSHEET GEOMECHANICS
AN INTRODUCTION

UNIVERSITY OF STRATHCLYDE

30125 00475582 2

ANDERSONIAN LIBRARY
★
WITHDRAWN
FROM
LIBRARY
STOCK
★
UNIVERSITY OF STRATHCLYDE

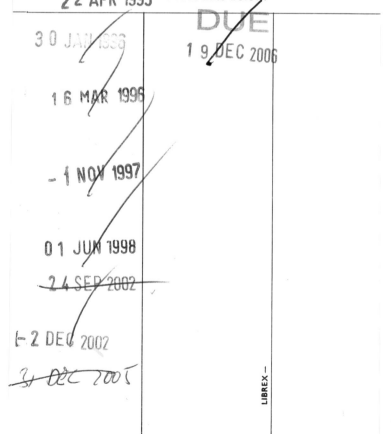

Books are to be returned on or before
the last date below.

2 2 APR 1995

3 0 JAN 1996

1 6 MAR 1996

DUE

1 9 DEC 2006

- 1 NOV 1997

0 1 JUN 1998

2 4 SEP 2002

1- 2 DEC 2002

3 DEC 2005

LIBREX —